P9-DEF-262

Bridges to Infinity

Bridges to Infinity

The Human Side of Mathematics

Michael Guillen

JEREMY P. TARCHER, INC.
Los Angeles
Distributed by Houghton Mifflin Company
Boston

Library of Congress Cataloging in Publication Data

Guillen, Michael.
Bridges to infinity.

Bibliography: p. 191
Includes index.
1. Mathematics—Popular works. I. Title.
QA93.G8 1983 510 83-15925
ISBN 0-87477-233-8

Requests for such permissions should be addressed to:
Jeremy P. Tarcher, Inc.
9110 Sunset Blvd.
Los Angeles, CA 90069

Design by Thom Dower

Manufactured in the United States of America

10 9 8 7 6 5 4 3

First Edition

To mom and dad, with love

CONTENTS

ACKNOWLEDGMENTS

Among those many friends and colleagues to whom I have incurred a significant debt of one kind or another in writing this book, I wish to thank most particularly Jane Livermore, Lars Wahlbin, Iain Johnstone, Richard Liboff, Warren Coates, Norma Bowles, Robin Raphaelian, Linda Greer French, Mary Nadler, and Joyce Boorn.

For meting out criticism and praise at just the appropriate moments along the way and for prodigious efforts on behalf of this book, special thanks go to my literary agent, Sallie Gouverneur, and to my editor, Janice Gallagher.

Most especially, I wish to thank Laurel Lucas, whose tender loving care, helpful comments, and extrahuman efforts in producing the manuscript enabled me to write a better book than I probably would otherwise have written.

Introduction
MATH ANXIETY

> Man consists of body, mind and imagination. His body is faulty, his mind untrustworthy, but his imagination has made him remarkable. In some centuries, his imagination has made life on this planet an intense practice of all the lovelier energies.
>
> John Masefield, *Shakespeare and Spiritual Life*

Most of the reasons that I have written this book can be explained in terms of what reportedly happened one day in the eighteenth century when the great German mathematician Leonhard Euler confronted the eminent French scholar and atheist Denis Diderot with a spurious mathematical proof for the existence of God. Euler, it seems, accepted an invitation to meet Diderot, who at the time was in attendance at the royal court of the Russian Czar. On the day of his arrival, the story goes, Euler strode up to Diderot and proclaimed: "Monsieur, $(a + b^n)/n = X$, donc Dieu existe; répondez!" [Sir, $(a + b^n)/n = X$, therefore, God exists; respond!] In the past, the French scholar had eloquently and forcefully refuted many a clever philosophical argument for the existence of God, but at this moment, at a loss to comprehend the meaning of this mathematical equation, Diderot was intimidated into silence.

The significance of this story is that it typifies the exchange between mathematicians and nonmathematicians in our society—which is really no exchange at all. My purpose here is not to blame anyone in particular as to why mathematics is widely misunderstood, but to remind us that it is misunderstood. Diderot's uncharacteristic dumbness in the face of Euler's proclamation is a reaction to mathematics that most people—including those as intelligent as Diderot—are able to identify with. It is the primary symptom of the age-old epidemic known today as math anxiety.

Math anxiety is the pathological dread and unabashed humility that mathematics evokes in hundreds of millions of people. Throughout history, most people have had this same reaction toward mathematics. But although the affliction itself has not changed that much, the consequences of being a victim of it *have* changed, and dramatically so.

For Diderot, being victimized by math anxiety meant that he was deprived of being able to consider for himself the manifold and unique ways in which mathematics bears on our far-flung human concerns, including our questions about God. Many of those connections are discussed in these essays.

For a person today, being similarly afflicted means what it did for Diderot, and more—it means being deprived of any intimate understanding of our complex technological world. And without such an understanding, a person is merely a spectator, rather than a participant, in the world.

It is difficult to overstate the value that numbers have in our descriptions of physical and metaphysical reality and to understate the incapacity of a person with math anxiety to fully appreciate those descriptions. Galileo once expressed the importance of numbers to the physical sciences by saying that the "book of nature" is written in the language of mathematics. The Cabala of Judaism is but one example of the enormous mystical significance most religions impute to numbers.

Math anxiety, like senility, is actually not just one but

several afflictions, each the outcome of some misperception about mathematics. The most important of these, illustrated in the Euler-Diderot incident, are ones that I expect will be readily recognized by any person who reads this book.

First and foremost, math anxiety is based on an unawareness of the overall limitations of mathematics. Doubtless, part of the reason that Diderot was so dumbstruck by Euler's amazing assertion is that he was not aware that mathematicians at the time had yet to come to terms with infinity, much less with God. As the essay "Beyond Infinity" explains, it was not until the late 1800s that the German Georg Cantor formulated the mathematical means to help us explicate the nature of infinity and of what lies beyond.

That sounds like an impressive achievement, and it was; it was also one of the first links in a chain of events, discussed in the first essay of the book "A Certain Treasure," that brought home to mathematicians some of the inherent shortcomings of their subject. Hitherto, mathematicians had believed in the limitless and infallible capacity of mathematics to define the truth logically. The shortcomings discovered have ultimately served to evidence the humanness of mathematics by revealing not only its fallibilities but also the persistence and optimism with which modern mathematicians are struggling to overcome the limitations.

The revelation of these shortcomings serves also to divide up the history of mathematics into periods that I refer to in this book as "fantasizing," "compromising," and "optimizing." These correspond, respectively, to the times before, during, and after the revelations that transpired between the mid-nineteenth and early twentieth centuries. In keeping with this distinction, the Fantasizing section of this book refers to those subjects that were largely elaborated at a time when mathematicians still believed in the infallibility of mathematics. The Compromising section includes a pair of essays that discuss the two main incidents that changed mathematicians' image of their subject. The Optimizing section is comprised of essays about topics that were largely formulated in this century, when mathematicians were already well aware of their subject's fallibilities and were trying to make the most of it. I invite the reader to see if he or she can detect in each essay the mathematicians' particular frames of mind.

Related in a way to an unawareness of the overall limitations of mathematics is a misunderstanding of what it is and is not; this misunderstanding, too, contributes to a person's math anxiety. Diderot would have had an easier time of it with Euler had he but understood that mathematics is not a science—it is not capable of proving or disproving the existence of real things. In fact, a mathematician's ultimate concern is that his or her inventions be logical, not realistic.

This is not to say, however, that mathematical inventions do not correspond to real things. They do, in most, and possibly all, cases. The coincidence between mathematical ideas and natural reality is so extensive and well documented, in fact, that it requires an explanation. Keep in mind that the coincidence is not the outcome of mathematicians trying to be realistic—quite to the contrary, their ideas are often very abstract and do not initially appear to have any correspondence to the real world. Typically, however, mathematical ideas are eventually successfully applied to describe real phenomena, as is discussed in some detail in the essay "Inventing Reality."

Most observers attempting to account for this remarkable concatenation of the mathematical and real worlds suppose, as Galileo did, that it is an indication of nature's mathematical basis. Because mathematicians speak nature's language, as it were, it would be natural from this explanation to expect that what they say will usually, if unintentionally, have some bearing on reality.

My own explanation of the coincidence supposes that the human imagination is, literally, a sixth sense. According to this explanation, mathematical ideas turn out to be realistic as often as they do simply because they are not merely inventions, but observations. We perceive reality with our imagination, I believe, in the same way that we perceive reality with our other five senses. The coincidence between the real world and the perceptions of this sixth sense is the same kind of coincidence that exists between the real world and our sensory perceptions of sight, sound, taste, feel, and smell.

Unlike scientists, who observe nature with all five senses, mathematicians observe nature with the sense of imagination almost exclusively. That is, mathematicians are as specialized, and therefore as well practiced, with this sixth sense as musicians are with sounds, gourmets are with tastes and smells, and photographers and filmmakers are with sights. This comparison also suggests that mathematicians are artists of the imagination just as surely as musicians, gourmets, photographers, and filmmakers are of their respective sensory domains. Through their unique creations, mathematicians inform us about reality without the intent, or ability, to actually prove that something does or does not exist.

The final significant misperception about mathematics illustrated by Diderot's reaction to Euler's cryptic claim is actually as common among mathematicians as it is among nonmathematicians. This misperception is that mathematics can be usefully expressed only in terms of symbols. This mistaken notion leads many mathematicians to dismiss as futile any efforts to explain mathematical results in ordinary language, and it leads nonmathematicians to lower their expectations about being able to understand mathematics. Indeed, it has long since come to the point where most people just automatically expect that they won't understand *anything* about mathematics.

Of all the aspects of math anxiety that I have described, this one in particular could easily be eliminated if mathematicians were to encourage a change in attitude. Sad to say, however, many of them believe as the twentieth-century British mathematician G. H. Hardy did. In his memoirs, *A Mathematician's Apology,* Hardy maintained that expository mathematical writing was an activity for second-rate minds. Persons who *do* mathematics, he wrote, despise persons who *explain* mathematics to others—presumably he considered such efforts to be a silly, foredoomed waste of time.

At the risk of establishing myself with my colleagues as a second-rate mind, I have written this book because I believe, as the eminent mathematician Bernhard Riemann did, that it is possible to convey the conceptual content of mathematics to a nonmathematical audience. On the occasion of his being considered for an academic position at the now famous Göttingen University in Germany, Riemann delivered a lecture on a most technical topic—the foundations of geometry—without resorting to a single equation. His lecture, which was attended by the university's entire faculty of learned men and women, was an enormous success.

It is with Riemann's example in mind that I have written these essays in plain English, without resorting to equations. This does not imply that the essays are written condescendingly; they are not. They are written so that Diderot, as well as Riemann, might have approved of my efforts, modest as they are.

Beyond providing the wherewithal to overcome math anxiety, these essays will, I hope, animate in you some of the same excitement about mathematics that I have felt ever since I was a child. In reading these essays, you will perhaps agree with me that the mathematician's imagination is as fruitfully applied to the questions of life as to the questions of arithmetic and geometry. If so, then I will have succeeded in challenging yet another misperception about mathematics.

The misperception is that mathematics is antiseptically rational and therefore has little or no relevance to the characteristically irrational activities and beliefs of human beings. This was a conviction expressed by Aldous Huxley when he wrote in *Views of Holland:* "We have learnt that nothing is simple and rational except what we ourselves have invented; that God thinks in terms neither of Euclid nor of Riemann. . . ." As I discuss in the essays "Between Checkers and Chess," "The Call of the Wild," "Abstract Symmetry," and "Nothing Like Common Sense," for instance, an understanding of mathematics can assist us immeasurably in our multidisciplinary attempts to understand human nature and human existence.

Huxley and those who agree with him have overlooked the possibility that mathematicians are not merely inventing ideas that are antithetical to the complexities of life; they might actually be looking at life with a most trenchant sense—one that perceives things the other five senses cannot. It was a sense that Diderot, for all his brilliance and insight, never practiced, although he *could* have. For this imagination of ours is a sense that can be cultivated; it requires no special intelligence, but only a desire to learn the ways of mathematics and mathematicians.

PART ONE

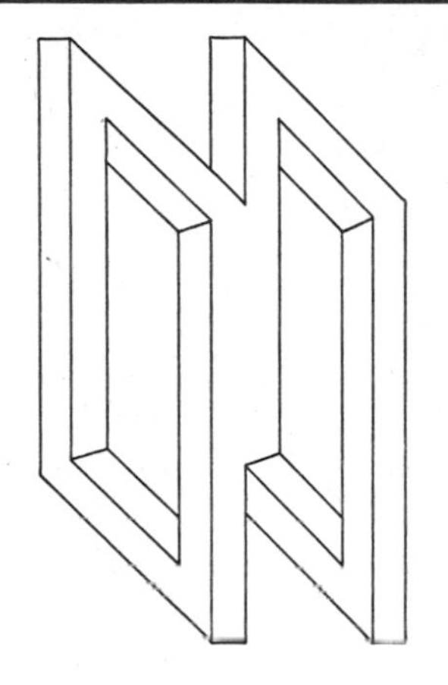

FANTASIZING

Logic and Proof

A CERTAIN TREASURE

> Proof is an idol before which the mathematician tortures himself.
>
> Sir Arthur Eddington

It is a rare person who does not prefer certainty over doubt in most matters and an even rarer person who can obtain it. It is as though certainty were a buried treasure, and we who so desire it have yet to find a map that can lead us to it. Around 300 B.C., mathematicians believed they had found such a map in the guiding principles of Aristotle's logic. Euclid followed those principles in proving the theorems of geometry (the study of shapes), which were hailed as models of certainty for the next 2000 years. During the late nineteenth century, however, when mathematicians guided by those same principles essayed to prove the theorems of arithmetic (the study of numbers), they were led to a treasure of paradoxes rather than of certainties. Aristotle's logic was flawed.

This revelation divided mathematicians into various schools of thought, with each group claiming to have the one map that would redirect all mathematicians to an infallible standard of proof. The divisiveness that ensued was superseded in 1931 by yet another revelation, which decreed that there is no conceivable way to obtain complete certainty in mathematics.

There was a finality to this edict that disheartened some mathematicians and inspired others to try to find some way around it—so far without success. Like the rest of us, most mathematicians have learned to accept doubt as a familiar part of doing business; it is a grudging acceptance, though, by those mathematicians who still nurse a hope of recovering the certainty that was once believed to be theirs.

The wave of mathematical certainty crested around 300 B.C., with the appearance of Aristotle's *Organon* (a Latin word meaning "instrument of reason") and Euclid's *Elements*. At that time it was widely believed that the *Organon* proffered the way to logical certainty and that the *Elements* was the treasure of indubitability itself.

In the *Organon*, Aristotle reduced the thitherto ill-defined process of deductive reasoning to fourteen rules and a few canons by which conclusions could properly be derived from assumptions. Among the canons, there was the law of identity (Everything is identical with itself), the law of contradiction (Nothing can both be and not be), and the law of the excluded middle (Something is either true or false; there is no third possibility). The canons were intended to express verities that most of us would call common sense, whereas the rules were the outcome of Aristotle's meticulous study of syllogisms.

A syllogism is a three-step exercise in deductive reasoning of the following form:

(if) All men are mortal
(and if) Socrates is a man
(then) Socrates is mortal

The first two statements are the assumptions, and the third statement is the conclusion, which, as Aristotle put it, "follows of necessity" from the assumptions. We may doubt the credibility of the assumptions, Aristotle explained, but if the rules of deductive reasoning have been followed, there is no doubting the conclusion. For example, one might dispute the assumptions

All happy people are amiable

and

Bill is a happy person,

but there's no disputing that from these assumptions it necessarily follows that

Bill is amiable.

This is precisely what is meant by logical certainty. Because each was stated in terms of an archetypal syllogism, Aristotle's rules of logic were widely perceived as guides to certitude. (Later, medieval logicians added five new rules to the original fourteen.)

In his magnum opus, the *Elements,* Euclid followed the principles of deductive reasoning in deriving hundreds of theorems in geometry from only ten assumptions. The assumptions were a mixture of common sense (Things equal to the same thing are equal to each other) and plausible assertions about mathematical points, lines, and planes (A straight line can be drawn from any point to any other point). Mathematicians could, and did, challenge some of Euclid's assumptions, but, as Aristotle had shown, there was no doubting the conclusions. Euclid's theorems were all *If . . . then* arguments derived according to the rules of logic. If a person believed in the infallibility of Aristotle's rules, he was bound to believe that Euclid's theorems were models of logical certainty, whatever he thought about the assumptions.

For more than 2000 years after the first appearance of the *Organon* and the *Elements*, not only mathematicians but also philosophers, scientists, and the literati believed that in Aristotle's logic and deductive reasoning they had a means by which to obtain surety in a wide range of matters. During the thirteenth century, for example, the Italian Scholastic Thomas Aquinas used Aristotelian reasoning to corroborate the veracity of matters of faith, including the existence of God. Thomism, as this synthesis of Aristotelianism and Christianity came to be called, was so influential that Pope Leo XIII issued an encyclical in 1879 declaring it the official philosophy of the Roman Catholic Church.

It was, therefore, just another ardent declaration of faith in the infallibility of deductive reasoning when mathematicians in the late 1800s embarked en masse on a program to do for arithmetic what Euclid had done for geometry. The general idea was to reformulate the hodgepodge of arithmetical results that had accumulated over the centuries into some kind of logical format. Mathematicians had come to accept many of these results without proving them, mostly because they seemed to be a matter of common sense. For example, no one had ever thought seriously to question the law of trichotomy, which states that every ordinary number is either zero, positive, or negative. The law was taken to be as indisputable as stating that every moment in time is either a part of the present, future, or past, and so no one could imagine it to be false.

It *isn't* false, as it happens, but as late as one hundred years ago, mathematicians had not yet gotten around to proving it, along with many other arithmetical truths. Only now was this long-standing nonchalance giving way to an eagerness to acquire logical certainty in matters arithmetic. These mathematicians expected that ahead of them was a short and rather routine quest, to be made following the same map that Euclid had followed. Actually, it was the beginning of a futile hunt for a nonexistent treasure.

Though several mathematicians led the rest throughout the time of the search, the German mathematician Gottlob Frege was one of the first to declare that he had finished it. He had worked from 1893 to 1902 to derive hundreds of theorems of arithmetic from just a few assumptions, and the tangible result was a monumental two-volume treatise entitled *Grundgesetze der Arithmetik* ("Fundamental Laws of

Arithmetic"). His assumptions, like Euclid's, might be challenged, but his conclusions were drawn according to principles of deductive reasoning that were consistent, though technically not identical, with Aristotle's. He and his contemporaries, therefore, had every reason to believe that his *Grundgesetze* was no less a model of certitude than the *Elements*.

They had no reason to believe otherwise, that is, until 1902, the year Frege was putting the finishing touches on his work. In that year, the English mathematician-philosopher Bertrand Russell made it known that he had spotted a paradox, a flaw in logic, in the final manuscript of Frege's second volume. The paradox was not due merely to some careless and easily correctable error on Frege's part, Russell explained; it was more in the way of a defect in deductive logic itself. No one at the time could possibly guess how serious a flaw it was or what it would require to eliminate it, but it was serious enough for Frege to recognize that it vitiated his entire ten-year effort. In a rather sad postscript to his second volume, Frege wrote: "A scientist can hardly meet with anything more undesirable than to have the foundation give way just as the work is finished. In this position I was put by a letter from Mr. Bertrand Russell as the work was nearly through the press."

Russell specifically prosecuted the apparently benign notions of class and class membership that Frege had used in his *Grundgesetze* to describe collections of numbers. According to Aristotelian logic, a class is any group of objects—automobiles, birds, or numbers, for example—that are related by certain qualitative similarities. Conversely, a class is defined by the very qualitative similarities of its members, just as a neighborhood is distinguished by the people who make it up. In order for a candidate to qualify for membership in a particular class, it must share precisely the qualitative similarities of existing members, however incompatible it may be otherwise.

Whereas other mathematicians of the time simply took these ideas for granted, Russell tested them in his mind. "It seemed to me," he recalled years later in *My Philosophical Development*, "that a class sometimes is, and sometimes is not, a member of itself. The class of teaspoons, for example, is not another teaspoon, but the class of things that are not teaspoons is one of the things that are not teaspoons."

Most classes we can think of are of the teaspoon sort. The classes of shoes, of houses, of pencils—none of these classes is a member of itself. Still, there are a few good examples of classes that *are* members of themselves. For instance, the set of all things printed on this page is itself printed on this page, and the set of all ideas is itself an idea. In each case, the class itself qualifies for membership because taken as a single entity it shares the qualitative similarities of its members.

Next, Russell recalled, he mused over the unimaginably huge class that contains all "the classes that are not members of themselves." Let's call this the NS class, for "nonself." The NS class includes all the familiar classes (the classes of teaspoons, shoes, and so on). Then, he wrote, "I asked myself whether this [NS] class is a member of itself or not."

It was in attempting to answer this question that Russell discovered the paradox. If we assume that NS is a member of NS, he reasoned, then we assume that NS is a member of itself. But, according to the original definition of NS, this very assumption disqualifies NS from membership in NS (by definition, remember, NS includes only classes that are not members of themselves). If, on the contrary, we assume that NS is not a member of NS, then we assume that NS is not a member of itself. But according to the original definition of NS, this very assumption decrees that NS qualifies for membership in NS. "Thus each alternative leads to its [logical] opposite," Russell wrote. "It put an end to the logical honeymoon that I had been enjoying."

Russell's paradox illustrated the fact that following the

rules of logic could lead us to contradictory results. An immediate implication of this was that Aristotelian logic had to be either somehow rehabilitated to remove the offending paradox or replaced with an entirely new means of obtaining complete certainty in mathematics. During the first three decades of this century, mathematicians disagreed about how best to set things aright (not knowing that an unexpected discovery would render all their differences of opinion moot). Of the many schools of thought that emerged in that thirty-year period, two of them were aimed specifically at rehabilitating Aristotelian logic. These were the Logicist and Formalist schools, and the programs they championed preoccupied the mathematicians of the time.

The Logicists, led by Russell himself, would have obviated Russell's paradox by modifying the rules for class membership. Specifically, they wanted to disallow a priori the possibility that a class could be a member of itself. To accomplish this, they proposed that the principles of Aristotelian logic be supplemented hereafter by what they called the "vicious circle principle," which states, "Whatever involves all of a collection must not be one of the collection." By the enforcement of this principle, the question of whether Russell's hypothetical NS class is or is not a member of itself would be settled by fiat, and the vicious circle of Russell's paradox would thereby be avoided.

In contrast to the Logicists, the Formalists believed that the shortcomings revealed by the paradox were not in logic itself but in the semantic content of the language used to express logic. In particular, they traced the origin of many paradoxes, including Russell's, to the ambiguous meaning of the word "all." Statements such as "All rules have exceptions" are either innocuous or paradoxical, depending on whether we interpret the word "all" to include or exclude the statement of which it is a part. Uncertainties such as these are semantic rather than logical, the Formalists maintained, and could be expurgated simply by bleaching logic of its semantic coloration. Thus, they set about re-expressing the logical arguments in mathematics in terms of strictly defined symbols that had no real meaning, rather than in terms of words.

Each of the routes described by these and other schools promised to lead mathematicians back to the treasure of certainty, and until 1931 each faction hopefully pursued its chosen way. But in that year all mathematicians were stopped dead in their tracks when the German logician Kurt Gödel announced his discovery that complete certainty was never to be encountered in mathematics by any route founded on traditional logic. It was an ironic proof in that Gödel had used logic in a clever way to establish its own shortcomings.

The gist of his finding was that any standard of proof based on the self-consistent principles of deductive reasoning is inadequate to establish the truth or falsity of every conceivable mathematical theorem, just as Aristotle's logic was inadequate to settle the question in Russell's paradox (that is, Is the NS class a member of itself?). In short, there will always be questions that arise in mathematics that cannot be settled with logical certainty. Furthermore, Gödel found, any imaginable remedy for an inadequate standard will also prove inadequate in the same way. This meant that the Logicists' and the Formalists' proposals were predestined to fail in the end.

Gödel's results stimulated the invention of non-Aristotelian logical systems, according to which a statement can be something other than true or false. The simplest of these is a so-called trivalent logical system, in which a statement can be either true, false, or merely possible. Such a system is based on a disregard for Aristotle's law of the excluded middle (that is, Something is either true or false; there is no third possibility), and because it allows for the possibility that a theorem might be logically uncertain, it is consistent with Gödel's finding.

For this reason, and because non-Aristotelian logical

systems are mathematically interesting subjects, some mathematicians today spend all their time studying and developing them. Efforts by other mathematicians also tend to be aimed at accommodating mathematics to, rather than extricating it from, Gödel's uncertainty. It is a tendency that is in marked contrast to the earlier fight to reclaim mathematical certainty from the throes of Russell's paradox.

In the aftermath of Gödel's revolutionary findings, most mathematicians resigned themselves to no longer thinking of mathematics as a bastion of certainty. One of these was the late Hungarian mathematician-philosopher Imre Lakatos, who in 1963 articulated a philosophy of mathematics that accommodated the Gödelian uncertainties. Lakatos described mathematics much as his mentor Karl Popper had described science, as always having a tentative status and as subject to being revised, even drastically, by new discoveries. Lakatos wrote: "Mathematics does not grow through a monotonous increase of the number of indubitably established theorems, but through the incessant improvement of guesses by speculation and criticism."

In 1981, the mathematician-historian Morris Kline was making much the same point in his book *Mathematics: The Loss of Certainty* when he compared the mathematician to a homesteader "who clears a piece of ground but notices wild beasts lurking in a wooded area surrounding the clearing." To increase his sense of security, the homesteader clears a larger and larger area, but he is never able to feel absolutely safe—"The beasts are always there, and one day they may surprise and destroy him." Similarly, Kline wrote, the mathematician uses logic to clear away a wilderness of mathematical ignorance, but he must expect that logical inadequacies may be discovered at any time that, like beasts, will spoil his hopes for complete security from doubt.

The mood in mathematics today, however, is not one of complete accommodation to Gödelian uncertainty. Perhaps it is because mathematicians coexist so uneasily with uncertainty that they go about their day-to-day business as though the events of this century had never happened. Or perhaps it is because, as Kline suggests, "they find it hard to believe that there can be any serious concern . . . about their own mathematical activity"—each mathematician behaves as if Gödel's uncertainty is something that affects the next person, but not him.

For whatever reason, mathematicians today "write and publish as if uncertainties were nonexistent," writes Kline. In practice, if not in principle, they retain the pre-Gödelian conviction that, as described by the German Formalist David Hilbert, "every definite mathematical problem must necessarily be susceptible of an exact settlement, either in the form of an actual answer to the question asked, or by a proof of the impossibility of its solution."

This adherence to a conviction that has been proved incredible is the human side of mathematics. It is not consistent with the popular image of a mathematician, but it is entirely consistent with human nature. In behaving as though certainty exists or can be had in mathematics, today's mathematicians are not unlike those many inventors who believe, against all odds, in the feasibility of a perpetual motion machine. Their conviction has been and continues to be in the nature of human progress, impelling one generation to prove possible what previous generations had proved impossible. In mathematicians, as in others, "this conviction . . . is a powerful incentive," Hilbert wrote in 1900. "We hear within us the perpetual call: There is the problem. Seek its solution. You can find it by pure reason, for in mathematics there is no 'we will not know.'"

In 1959, a disillusioned Russell lamented: "I wanted certainty in the kind of way in which people want religious faith. I thought that certainty is more likely to be found in mathematics than elsewhere. . . . But after some twenty years of arduous toil, I came to the conclusion that there was nothing more that I could do in the way of making mathematical knowledge indubitable."

Doubtless there are and will be many more mathematicians like Russell who will dedicate decades of their lives to the search for the treasure of certainty. So long as this remains the case, there is always the possibility that one of them will encounter it and thereby bequeath to future generations what previous mathematicians have failed to: an infallible standard of proof.

Related Essays

Much Ado About Nothing

An Article of Faith

Limit and Calculus

LOCATING THE VANISHING POINT

> The will is infinite
> and the execution confined,
> The desire is boundless
> and the act a slave to limit.
>
> Shakespeare, *Troilus and Cressida*

We are a singularly restive species, driven by some natural tendency to want to outdo ourselves. Perhaps it is but an expression of the universal tendency that Darwin described when he wrote, "As natural selection works solely by and for the good of each being, all corporeal and mental environments will tend towards perfection." Or perhaps it is, as the French philosopher J. B. Robinet suggested, a tendency unique to the human mind, whose "destiny can be nothing other than to exercise imagination, to invent, and to perfect." In either case, it appears certain that what we are tending toward—call it perfection, if you like—is an unattainable goal. At least, it is not *likely* that we will one day reach perfection—if we could, in fact, recognize it—and then strive no longer. Rather, perfection in our psychology is like the vanishing point in a painting, an imaginary point at infinity toward which every major course is directed. Most especially, it is like the mathematician's asymptotic limit: an ideal that is nameable and forever approachable yet never realizable.

In mathematics, the asymptotic limit is an important concept. It is one that I still find incredible, even after having been exposed to it for so many years. I should not be surprised at my difficulty with it, however, because the idea of an asymptotic limit is inextricably tied to that of another obdurate enigma—infinity.

This is especially apparent in geometry, where the concept of an asymptotic limit is portrayed by an infinitely long boundary line called an asymptote. By definition, another line may approach an asymptote with increasing proximity but never actually meet it, like an airplane that approaches a runway but never actually touches down. I keep wanting to imagine that such a line will eventually close the gap between itself and the asymptote, but this is not so. As the line and asymptote converge, the distance between them is being continually halved, but no point is ever reached where the remaining half vanishes completely—not even at infinity.

This is precisely what I find so incredible about asymptotic limits. In discussing them, mathematicians are able to tell us literally about how some things behave at infinity. Admittedly, with some examples, this is not really much of a feat; the asymptotic limits of some things are readily guessed at. Imagine, for instance, a many-sided figure, a polygon, inscribed in a circle. Next, imagine systematically increasing the number of sides of the polygon. Although this is an interminable process, I think it soon becomes obvious that the polygon is tending to become a circle, its asymptotic limit. Similarly, I think it's clear that the asymptotic limit of the numerical sequence .9, .99, .999, .9999, and so on, is simply the number 1. It is never *actually* 1, but it is forever becoming a number that comes closer and closer and closer to 1.

With most examples, however, behavior at infinity is not so obvious, yet even in these cases mathematicians have discovered ways of actually determining the asymptotic limit.

Most of these methods are based on the ability to add up an infinite sequence of numbers—something which may sound impossible, but at which mathematicians have become quite adept during the last 300-odd years. Using this skill, they know, for instance, that 1 plus ½ plus ¼ plus ⅛ plus 1/16 plus 1/32 and so forth, ad infinitum, eventually adds up to 2—not approximately 2, mind you, but *exactly* 2.

This skill is very handy in predicting asymptotic limits, since they are by definition the results of interminable progressions. For example, imagine a three-sided polygon, a triangle, inscribed in a one-inch circle. Next, imagine inscribing a circle within this three-sided polygon, and then a four-sided polygon within this circle, and then another circle within the four-sided polygon, and so on. Imagine continuing this process indefinitely, always following an inscribed circle with an inscribed polygon that has one more side to it than the last one had. Clearly, the inscribed figures will get smaller as you proceed.

You might guess that the limit of this interminable process is a point at the very center of all these circles and polygons, but that is not so. Remember that the increasingly faceted polygons will tend to become circles themselves, which means that the nesting process at some point almost amounts to laying circle upon circle. More precisely, from some point onward there is a vanishing difference between the shapes and sizes of the circles and polygons. The net effect of this is that the interminable nesting of figures tends toward a small circle that is concentric with the original circle. Using their skill with infinite series, mathematicians are even able to predict the diameter of this limiting circle—this unrealizable ideal, as it were—to be roughly 1/12 of an inch. With an analogous skill, we would be able to divine precisely which state of perfection (or imperfection) human beings are perpetually approaching, or which record time for running the one-mile race will always be approached but never actually reached.

One of the biggest payoffs of mathematicians' skill with

asymptotic limits has been a prodigious theory called the differential calculus, invented by Isaac Newton and Gottfried Leibnitz in the seventeenth century. This theory is useful for describing, in perfect detail, smooth-going changes of almost any kind. By "perfect detail" is meant that with the differential calculus it is possible to codify each and every instant of change in a dynamic process. NASA scientists use it to calculate the instant-to-instant progress of a satellite bound for Saturn, and economists use it to monitor the instant-to-instant progress of market trends.

Before the seventeenth century, mathematicians and scientists had had to settle for descriptions of change based on averages taken over measurable intervals of time. Typically, such descriptions are far from ideal. Suppose, for example, that you wanted to determine the yearly pattern of rainfall in the wettest spot in the United States, Mt. Waialeake in Kauai, Hawaii. According to your almanac, it rains there an average of 486.1 inches every year. The yearly average alone tells you nothing precise about the actual pattern of rainfall, with all its short-term and long-term fluctuations. Clearly, you would know more about this pattern if you knew how much it rains from one day to the next, or better still, from one hour, one minute, or one second to the next. Ideally, you would know all there is to know about the pattern of rainfall on Mt. Waialeake if you could but determine how much it rains from one instantaneous moment to the next—and it is the differential calculus that could give you the technical means to do this.

The invention of the differential calculus was based on the recognition that an instantaneous rate is the asymptotic limit of averages in which the time interval involved is systematically shrunk. This is a concept that mathematicians recognized long before they had the skill to actually compute such an asymptotic limit. For instance, they could calculate an employee's hourly wage; they merely had to divide the weekly salary by the number of hours in a work week. To calculate the employee's wage per minute, they had to divide the weekly salary by the number of minutes in a work week; they could proceed in this fashion interminably, generating a sequence of averages with successively smaller time intervals. But they didn't know how to actually calculate the asymptotic limit of such a sequence. According to common sense, the limit to the sequence is simply the weekly salary divided by the number of infinitely small time whits in a work week. But that number is infinity, and a sum divided by infinity is zero, which is a nonsensical result. It implies that an employee making $200 a week does so by earning *nothing* at any and every given instant on the job. Without a clear understanding of infinite sequences, precalculus mathematicians had as much chance of resolving this paradox as a nonmathematician would have of computing the correct asymptotic limit of the infinite nesting of circles and polygons. With the differential calculus, they gained the technical means of properly ascertaining the asymptotic limit of an average rate—that is, the instantaneous rate—and therefore the means to describe patterns of change with perfect precision.

This accessibility of mathematicians to perfection reminds me of a postcard I once received. The picture shows a pair of railroad tracks in perspective, with a group of jocular little men peering through magnifying glasses at the point on the horizon where the tracks appear to converge. The postcard was entitled "Locating the Vanishing Point." With the technical skills for computing asymptotic limits, mathematicians seem able to locate vanishing points. If we did not have these skills, such ideals would always remain beyond our reach, much like mountain ranges that appear to stay on the horizon no matter how much progress we make in their direction.

For now, we can only fantasize about what it might be like to locate perfection in our nobler social and personal efforts as readily as we now calculate instantaneous rates or any of the other asymptotic limits in mathematics. One of the difficulties is that our progress in such nonmathematical efforts is not usually like a sequence that has any logical pattern to it, in contrast to what is always the case with mathematical examples. Even if each of us had a clear idea of the meaning of progress and perfection, it is unlikely that all our ideas would converge to a single asymptotic limit. In this sense our natural tendencies cannot be likened to straight lines directed toward a single vanishing point; actually, the lines are crooked, many are aimless, and the point is a spot—a rather large, diffuse one.

There are instances outside of mathematics, however, where it would seem that the asymptotic limit ought to be as well-defined as it is for, say, the sequence .9, .99, .999, and so on. These are instances of relatively deterministic behavior in which everyone involved is intent on the same goal, and their collective progress is as sure and steady as a sequence of numbers. This is true with most athletic competitions. In the one-mile run, for example, you might expect that there is a fastest possible running time.

Years ago, everyone was guessing that the asymptotic limit for running the one-mile race was four minutes. No one, it was believed, would ever run a four-minute mile. That was before 1954, when Roger Bannister of England ran the world's first sub-four-minute mile in the record time of 3 minutes, 59.4 seconds. After that, people were guessing that no one would ever crack the 3-minute, 55-second barrier, but then in 1958 Herb Elliot of Australia did, running the mile in 3 minutes, 54.4 seconds. In fact, I've charted the progress of world record mile times set since 1865, and I find that the rate of improvement today is roughly the same as it has been for the past forty years; the record times for the mile run don't seem to be leveling off and approaching any predictable asymptote. The latest record time, set in 1981 by the Englishman Sebastian Coe, is 3 minutes, 47.33 seconds—well under four minutes.

Unless we're willing to believe that one day some miler

will actually set an absolutely unbreakable record, we must assume that the limit to our improvement in the mile run is an asymptotic limit. And with that knowledge there arises the question of what the exact nature is of a force or resistance that could keep us from reaching some absolute limit, even as it allowed us to approach it with greater and greater proximity. Presumably, the source of the resistance is related somehow to the very real energy and strength limitations of the unassisted human body.

To get an idea of how this mechanism must behave, imagine for the moment that the asymptotic limit in question is none other than the speed-of-light barrier (a miler going at that speed would take only about 5.37 millionths of a second to finish the race—about as long as it takes a person to react to an unexpected stimulus). In that case, the resistance mechanism is explained in the theory of special relativity.

According to this theory, by the time any material object gets up to about 9/10 the speed of light, an enormous amount of energy is required to increase its speed by even a small amount. The physical reason for this is that as the material object nears the speed of light, more and more of the energy intended for its propulsion inevitably gets converted into mass, which weighs down the object; the object then requires more propulsion to keep it going; the added propulsion gets converted into more mass, and so forth. Exactly how or why this happens is not clear, but we are certain that it does happen. (Electrons travel around a cyclotron at speeds approaching .999 that of light speed, and physicists have observed for years that such electrons behave sluggishly, as though they were many times their normal mass—just as the theory predicts.) The increasing mass of an accelerating material object combines with the diminishing effectiveness of its propulsion to make it physically impossible for the object to reach the speed of light—though, with extraordinary effort, it can come arbitrarily close.

Hypothetically, at the asymptotic limit—at the speed of

light—an object would be infinitely massive, and all of its propulsion energy would merely go to increasing that mass. Clearly, however, even under unrealistically ideal running conditions (no air resistance, for example), the human body by itself does not have the energy to come anywhere near the speeds where these relativistic effects are noticeable nor does it have the strength to endure the rigors of such high speeds. Typically, each of us consumes a few thousand calories of energy per day; it would take something like 6000 calories just to get us up to sixty miles per hour. Certainly, then, if there is an asymptotic limit to our running speed in the mile race—or in any other race, for that matter—it lies far short of the speed of light, but it is enforced by some other, similarly peculiar resistance mechanism.

Probably the most reliable way of estimating the limit is to wait until the sequence of world record times begins to level off. This has not happened yet, but when it does we'll know, because new records will be set less and less often and will improve on the old records by smaller and smaller amounts. At that point, it will become more obvious which record time our runners are probably zeroing in on.

Without certain knowledge of this limit, however, we will probably behave then just as we do now: as though there is no limit. That is precisely the attitude one would expect from a species whose progress is modeled after an asymptotic limit. The chief paradoxical characteristic of any approach toward an asymptotic limit is that the future promises to be an interminable sequence of improvements; it appears that when we reach one horizon, there will always be another. Only a certain knowledge of the asymptotic limit, gained by some prescient means, could possibly vitiate this illusion, but that for now is not an eventuality that a singularly restive species need worry about. At the moment, we are not even close to acquiring a means analogous to the differential calculus that would enable us to locate the vanishing points of human progress.

Continuity and Numbers

IRRATIONAL THINKING

> That vast ideal host which all His works
> Through endless ages never will reveal
>
> Mark Akenside, *Pleasures of Imagination*

The world, it appears, is going digital. There are digital recordings, digital images, digital computers, and digital watches. I find this last trend especially engaging in light of the age-old question of whether time flows continuously, or whether it proceeds haltingly in discrete ticks called chronons. Today's digital watches promote the latter impression, but the truth is that we are not any closer to settling the question than were the ancient Greeks when the horological rage was the clypsedra, a water clock that promoted the continuous flow idea of time. If anything, the question is more daunting today than ever; while scientists have been wondering whether or not time is continuous, their mathematical colleagues have changed the working definition of continuity from one that was scientifically sensible to one that is apparently scientifically imponderable. Now only time itself can tell us whether scientists will stay with the original mathematical notion of continuity or will adopt the modified one for the purposes of studying time.

The main reason scientists are concerned with what mathematicians have to say in the matter is that the quantitative model of continuity is a number line. This is an ordinary line along which mathematicians arrange the numbers known to them from a study of arithmetic; in effect, it is the mathematician's conceptual ruler. And in that context, the quantitative notion of continuity is represented by a number line whose each and every point is labeled with a number, such that it has no unlabeled gaps.

The most primitive number line imaginable is not continuous. It consists only of the whole numbers 1, 2, 3, 4, and so on, which, when placed at regular intervals along its length, leave obvious gaps. The first continuous number line was that of the ancient Greeks. In their number line, the gaps between whole numbers were interminably subdivided into fractions, apparently leaving no point unlabeled. It was named the rational number line, because together the whole numbers and the fractions constituted what the Greeks called rational numbers; these were numbers that could be expressed as the ratio of whole numbers (3:2, 2:7, 5:1).

One particular Greek who believed not merely mathematically but religiously that the rational number line represented the ideal model of continuity was Pythagoras of Samos. He was one of a cabalist cult who saw divinity in the image of continuity and especially in the numbers used to define it. He and his fellow cultists placed a great deal of importance, for instance, on the Greeks' discovery that musical octaves are produced by plucked strings whose lengths are in ratios of whole numbers. To the Pythagoreans this was a sure sign that everything about and within the universe, physically and metaphysically, could be described in terms of rational numbers. Their worship of numbers led them to develop a complex numerological iconography, the basis of modern numerology, in which the number 1 was the divine creator (presumably because it is the very first nonzero number). The numbers 2 and 3, for reasons known only to the

Pythagoreans, stood for femininity and masculinity, respectively; from this they surmised that the number 5 (femininity plus masculinity) stood for marriage.

Because there was no evidence to dispute it, the Greeks adhered for years to the rational number line model of continuity. It was not until sometime during the sixth century B.C. that the Pythagoreans, of all people, discovered holes in the model and, by inference, holes in their deity. The revelation came unexpectedly during their efforts to solve a rather ordinary problem. It was a problem similar to the question, How long is a fence that cuts a one-mile square plot of land diagonally into two triangles?

Normally, the Pythagoreans would have solved such a problem easily by using their own Pythagorean theorem (the same one misstated by the scarecrow just after receiving a university diploma in the 1936 movie version of *The Wizard of Oz*). According to this theorem, the length of any third side of a square-cornered triangle can be calculated by knowing the lengths of the other two sides.

In the past, the theorem had been used to find the exact lengths of many a square-cornered triangle. However, much to the consternation of the Pythagoreans, in this case their theorem was unable to proffer an exact numerical answer to the question. Their calculations could only tell them that the fence was almost, but not exactly, 3/2 miles long; the exact number was not in their mathematical vocabulary. It was, in two senses of the adjective, an *irrational* number: literally, because it wasn't any ratio of whole numbers, and figuratively, because it belonged nowhere in the conceptual realm of the ancient Greeks. As the Pythagoreans were to discover, it was also merely representative of many other numbers just like it.

The discovery of irrational numbers, which the Pythagoreans considered heretical and apparently tried at first to conceal, meant that the rational number line was not a model of continuity after all. For instance, imagine a stick as long as the diagonal of that one-mile square in Pythagoras's problem. The stick has a length whose reality cannot be denied, yet when it is measured against the rational number line its end falls next to an unlabeled point. Furthermore, there are many such sticks, whose lengths can be expressed only in terms of numbers that are neither whole numbers nor fractions. In short, the Pythagorean discovery meant that the rational number line is riddled with gaps that in some ineffable way exist between its already infinitely close fractions.

In the centuries that followed that discovery, irrational numbers came to be accepted by mathematicians as a necessary evil. They were considered necessary, because by their fitting somehow into the discontinuities of the rational number line they created a more genuinely continuous number line; and they were considered evil because it was as yet unclear just how they fitted into the rational number line.

There were many examples of irrational numbers, but there was no acceptable mathematical definition of them. They appeared to be unrelated logically to the rational numbers, which is to say that there was no known recipe according to which an irrational number could be made from a rational number. By contrast, the rationals themselves *were* related logically to one another by dint of their all being ratios of whole numbers. As long as mathematicians were unable to reconcile the irrationals with the rationals, the concept of an irrational number line could not be considered fully understood, let alone validated mathematically.

Even well into the nineteenth century, a mathematician trying to rationalize irrational numbers was like someone at a family reunion who has not figured out just how some strange-looking person fits into the family pedigree. In 1872, the German mathematician Richard Dedekind found an acceptable way of relating the irrationals to the rationals and thereby of christening the long-anticipated "super" model of continuity. It was named the real number line, in reference to the habit of mathematicians since the sixteenth century of calling the rationals and irrationals together the real numbers.

A few years after Dedekind's achievement, the German mathematician Georg Cantor made a discovery that would lay to rest any lingering skepticism toward the seemingly impossible constitution of the real number line. He confirmed that though there are an infinite number of rational numbers, as the Greeks had suspected, there are even more real numbers. This confirms that the real number line is indeed somehow more continuous, more densely packed, than the ponderably continuous rational number line.

Imagine what this means: On the rational number line, adjacent numbers are infinitely close together. But with the real number line, adjacent numbers are *closer* than infinitely close together. Even using mirrors, this would be an impossible illusion for us to create. Consider, for instance, two mirrors facing each other. If you were to look in one of the mirrors, you would see a repeated image of the other mirror. Imagine that each image is a tick mark on a number line. If you were to bring the two mirrors closer together, the repeated image would be compressed and the spacing between them would shrink. The spacing between ticks in a rational number line would correspond to the spacing between images when the two mirrors were infinitely close together—touching, in other words. The spacing between ticks in a real number line, however, would correspond to the spacing between the images when the two mirrors were *closer* than touching. But this could only happen if the mirrors actually entered one another, as Alice entered the looking glass.

For scientists interested in learning about time, the imponderability of the real number line poses an interesting choice. They can reject it out of hand for its being an unscientific (that is, not mensurable) model of reality, in which case they might be faulted by some for being narrow-minded. Or, they can accept it as a possible model of reality and be faulted by some for being mystical.

No unanimous choice is liable ever to be made, so let us assume for the moment that both the rational and real number lines are, each in its own way, viable models of continuity. In this case, the question we are considering is not only whether time is continuous or not, but whether, if it is, it is "rationally" continuous or "really" continuous.

As to the first of those questions, if time is not continuous, then it means that the existence of everything in the universe, including the universe itself, is played out as if it were on film, frame by frame. It means that temporal existence is a succession of momentary ticks, and that the appearance of continuity is merely an illusion, like that created by modern motion pictures. In the case of a motion picture, which is projected at twenty-four frames per second, the illusion works because our eyes retain each frame's image long enough for it to appear to melt smoothly into the next one; the discontinuities pass unnoticed. If the images were projected at too slow a rate, or if our eyes didn't naturally retain images, then we would see the discontinuities, and motion pictures would appear to flicker.

In the case of the universe, experiments using scientific instruments that act like eyes with very little retention have yet to find that physical reality flickers. If time *is* discontinuous, the results of these experiments imply that the projection rate of reality must be faster than about one hundred billion, trillion (a 1 followed by twenty-three zeros) time frames per second.

If time is continuous, however, then there is the question of whether our temporal existence has the continuity of a rational or of a real number line. Each model is impossible to verify by direct measurement, since in a manner of speaking it stands for a projection rate of reality that is either an infinite number of time frames per second, or faster. The most that any direct measurement can ever do is to show that the projection rate of reality must be greater than so many time frames per second; it can never find that the rate is actually infinity, let alone greater than infinity.

Indirect experiments, however, might give us a clue. One simple experiment that comes to mind is taking an inventory of various time-related natural phenomena to see whether rational numbers alone are sufficient to describe them quantitatively. If they are, this may be a clue that time and irrational numbers have nothing to do with each other, and that by implication the rational number line alone is an adequate and accurate model of temporal continuity. If they are not, however, then maybe that, too, is a clue that—as the Pythagoreans discovered once before in another context—the rational number line is an inadequate model of temporal continuity.

For this proposed inventory, it will be necessary for us to distinguish a rational number from an irrational number. The most recognizable aspect of an irrational number is that it can be expressed only as an infinitely long string of digits without any apparent pattern. Two examples are the square root of two (1.414213562 . . .) and pi, the number always obtained by dividing the circumference of any circle by its diameter (3.141592654 . . .). Pi is one of those especially popular irrational numbers that has been calculated explicitly out to hundreds of thousands of decimal places without there being any discernible pattern to its digits, as expected. (This is one thing I find incomprehensible about an irrational number: Its digits are arranged randomly, but not arbitrarily. If I were to interchange just one pair of digits in its infinitely long random string of digits, I would no longer have the same irrational number. Each irrational number has its unique randomness and each of its digits has a unique role in creating that individualized randomness.) A rational number, too, can be written in decimal form as an infinitely long string of digits, but by contrast there is always a discernible pattern ($2/35$ is 0.05714285714285714 . . . , $1/99$ is .0101010101 . . . , and 6 is 6.0000000 . . .).

Certainly there are many temporal phenomena known to us today that are perfectly describable in terms of rational numbers alone. I think immediately of electricity, the movement of electrical charges in a wire. Because of a Nobel Prize–

winning experiment by the American physicist Robert Millikan in 1906, we know today that electrical charges exist only as whole-numbered multiples of an indivisible elementary charge. Actually, there are two kinds of elementary charges; they have the same strength, but one is positive and the other is negative, in such a way that like charges repel and unlike charges attract. There is some speculation these days that there are subnuclear particles, called quarks, that tote electrical charges that are one-third and two-thirds of an elementary charge, but so far there is no suggestion, or evidence, that anything exists with an irrational number for a charge.

Then there is chemistry, in which the dynamics of a reaction are determined in large measure by the properties of the chemical elements involved. Those properties, in turn, are associated with the atomic numbers of the elements.

The atomic number of an element is equal to the number of electrons in each of its atoms, and it is always a rational number—a whole number, to be precise. For example, elements with atomic numbers 2, 10, 18, 36, 54, and 86 are chemically inert; they don't react at all with the other elements. By contrast, elements with atomic numbers 9, 17, 35, 53, and 85 are the halogens—they are very reactive, especially in producing salts. Of particular importance to us and to other organic life forms is the number 6, which stands for carbon. Carbon's unique affinity for bonding with other elements causes it to form long chainlike molecules like those of amino acids, proteins, and DNA.

In biology, too, many temporal phenomena are describable in terms of rational numbers. Each plant and animal species, for instance, appears to be characterized by a unique number of chromosomes in its individual cells. And this chromosome number, like the atomic number, is always a whole number.

Despite this preeminence of rational numbers, science

does need irrational numbers. For well over a century, scientists have been taking note of a growing inventory of special quantities whose appearance in nearly every scientific theory signifies their import in the modern description of space-time. These natural constants can be seen as nature's vital statistics, and right now it looks as though every one of them is an irrational number. For example, one of these constants, the speed of light, has been measured out to nine decimal places, and the digits have yet to show any pattern. (Expressed in millions of meters per second, our best measurement of the speed of light is the number .299792458.) Another constant is one that is descriptive of dynamic behavior at the atomic level. It is called the fine-structure constant, and there is no pattern to its digits even when measured out to ten decimal places. (Our best measurement of the fine-structure constant, which is a dimensionless quantity, is .0072973503.) In physics alone there are more than a dozen of these constants, which have been measured out to anywhere from a few to eleven decimal places, and not one of them has a pattern to its digits.

This evidence for irrational numbers in time-related natural phenomena would fulfill the intention of our experiment on temporal continuity where it not for one remaining difficulty: A peculiarity of any attempt to measure an irrational number is that it can not really be *measured*. In measuring a number such as the fine-structure constant, we can never totally eliminate the possibility that a pattern will emerge at some point in the measurement's growing string of digits. A decade from now, we will perhaps have determined that the fine-structure constant is, say, .00729735037297З, signaling us with its pattern that it might be a rational number after all. Such a pattern in the digits would have to persist indefinitely for the constant to be adjudged a rational number, however, and that is something we could never verify.

What this all comes down to is that in merely discovering irrational numbers, mathematicians have raised a scientific possibility that cannot be settled by any imaginable measurement or series of measurements. Although there is, mathematically, a rational distinction between time that is "really" continuous and time that is "rationally" continuous, it is a distinction that for the empiricist is altogether irrational.

Related Essays

Beyond Infinity

An Article of Faith

Cantorian Set Theory and Transfinite Numbers

BEYOND INFINITY

Du glcichst dem geist den du degreifst.
[You resemble the thought which you conceive.]

Goethe

I marvel at the human imagination, which is so vivid that it often drives us to debate the fine points of our fantasies. In the past, intelligent men and women have disagreed about the number of angels that would fit on the head of a pin; today, military experts who believe we can endure a global nuclear war quibble about the relative merits of postwar scenarios. It is as though we become at times like Walter Mitty, so seduced by a conceptual possibility that our normal, practical sense of reality is left behind.

In mathematics, this power of the human imagination to engage us in highly detailed reveries is constantly being tried to its outermost rational limits. And in their modern attempts to rationalize, to particularize, one of the most ineffable concepts of all time, infinity, mathematicians have had unusual success.

For millennia, up until the nineteenth century, the idea of infinity was anything but well-defined. For the spiritually inclined it had vague, theological connotations, and for others, including scientists and mathematicians, it was a catchall for what lay beyond the limits of rational thought. In this awestruck regard for infinity, we were, as George Bernard Shaw wrote in *Man and Superman,* like the bushman who cannot count further than his fingers: "To him . . . eleven is an incalculable myriad."

By the seventeenth century, scientists and mathematicians had begun to speak as though infinity might some day become a rational concept, but infinity was still a generic term for them, used to describe anything and everything incomprehensibly large. There was no distinguishing infinities. For example, in his *Dialogues Concerning Two New Sciences,* Galileo expressed his belief that a three-inch line contains just as many points—an infinite number—as a line twice as long. He accepted the paradox as might a person for whom the universe would be no more or less accessible were it one-half or twice as large.

A century after Galileo, infinity paradoxes such as these had only increased in number and severity. In 1851, in a little book entitled *The Paradoxes of Infinity,* the Czech mathematician Bernhard Bolzano attempted the first thoroughly rational treatment of the subject, but as a result really left us, if not further from, then certainly no closer to infinity than we had been before.

If mathematicians today are as close to comprehending infinity as anyone ever has been, it is chiefly due to the trenchant imagination of Georg Ferdinand Ludwig Philipp Cantor, a soft-spoken German math professor who was but six years old when Bolzano's book appeared. Cantor didn't necessarily intend to reason his way to infinity. Like other mathematicians during the last quarter of the nineteenth century, he was motivated by recent discoveries that impeached the veracity of geometry as the foundation of mathematics. Mathematicians were then in an intellectual panic, working to replace Euclidean geometry with a new cornerstone; this time, many believed, it would have to be arithmetical, not geometrical. It would have to be a rational framework based conceptually on the whole numbers, fractions, and irrational numbers (decimal numbers, such as pi, that are not fractions), and not on points, lines, and planes. It was a broadly, even vaguely, stated goal, and the different tacks taken by mathematicians toward it led them in various directions. Cantor's tack led him to infinity and, what was most unexpected, even beyond it.

Cantor began this far-flung journey of the mind by imagining a finite "collection . . . of definite and separate objects of our intuition or thought," which he called simply a finite set. With the ostensible goal of inventing an arithmetical foundation for mathematics, he intended to consider sets specifically of numbers, but his definition referred just as properly to sets of calendar months, of people, or of anything else.

According to Cantor's logical framework, one set would be called equivalent (in size) to another if the elements of one could be paired numerically with those of the other. By electing to define equivalence in just this way, Cantor was preparing to make it easy to compare really large sets, such as the set of seats in the Los Angeles Coliseum and the set of spectators who show up to see some event there. Following Cantor's definition, the way to find out whether these two sets are equivalent would be to see what happens when everyone sits down, or tries to. If some spectators are left standing, or if some seats are left vacant, then the two sets are not equivalent; otherwise, they are.

Cantor's definition does not require us to count, or even to know, the population of either of the two sets in order to assert their equivalence. This is ultimately what made it feasible for Cantor to compare and discriminate rationally between infinities, or, more precisely, between sets of infinitely large populations.

With just these few primitive ideas behind him, Cantor took his first step toward infinity. He illustrated how any finite set could be used as a stepping stone to define another, larger, finite set, and so forth, ad infinitum. At each step, the larger set is comprised of all the subsets that it is possible to cull from the set before it. According to his arbitrarily chosen use of the term, Cantor explained, a subset will be allowed to consist of none, any, or all of a set's elements. The subset with no elements will always be called the null subset, and the one with all of a set's elements will be called an improper subset. The other subsets of a set will be referred to as proper.

Any set containing one or more elements would be a suitable starting point for Cantor's stepping-stone argument. For instance, a two-element set (A, B) defines a four-element set, comprised of its four subsets (A, B, AB, and the null subset). A four-element set (A, B, C, D) defines a sixteen-element set (A, B, C, D, AB, AC, AD, BC, BD, CD, ABC, ABD, ACD, BCD, ABCD, and the null subset), and so forth.

We can even predict the exact numerical leaps between Cantor's stepping stones using a simple formula that was already known during his time. According to this formula, the number of subsets we can create from a finite set with x number of elements is "2 multiplied by itself x times." In symbols, the phrase in quotes is normally written 2^x. Thus, in symbols, a two-element set has 2^2 (that is, four) subsets and a four-element set has 2^4 (sixteen) subsets.

If Cantor had stopped after having defined his stepping stones, then he would have provided us with a well-organized, rational prescription for getting to, but not necessarily reaching, an infinite set. His work would have endeared itself to those who argued then and who argue today that infinity is like a verb rather than a noun, that infinity is implied by a limitless process and is not something actual that can be labeled. As it is, after laying his stepping stones, Cantor proceeded to posit the existence of a veritable infinite set.

Apropos of his goal of assigning the principal roles of his logical framework to numbers, Cantor took as his definitive example of an infinite set the whole numbers 1, 2, 3, 4, and so on. In his theory, in other words, the whole numbers were to be treated not as an interminable sequence, a verb, but as an actual infinite set, a noun. It was as though in his view of infinity Cantor had been inspired by the image evoked by the line in a poem by William Blake, ". . .[to] hold infinity in the palm of your hand."

In the course of his own poetic reverie on the subject, Cantor didn't merely name the existence of an actual infinite set, he also described it as, paradoxically, the only conceivable set that could be equivalent to parts of itself.

To illustrate this description, Cantor turned once again to the whole numbers. Consider, he said, that every other number in the infinite set of whole numbers is even. You might guess that the set of even numbers, albeit infinite, is only half as large as the infinite set of whole numbers. Such a guess conforms to the logic of our everyday, finite spheres in which the whole is equivalent to the sum of its parts. However, following Cantor's definition of equivalence, the infinite set of even numbers is exactly equivalent to the entire infinite set of whole numbers.

Cantor could not prove the equivalence by literally pairing each even number with each whole number, of course, but he did assert it as an indirect proof that it was inconceivable how such a pairing could possibly leave an excess of either kind of number, since each set of numbers was inexhaustible. For him, if not for many of his contemporaries, the paradox that a whole could be equal to one of its parts was an inescapable, rational trait of the infinite realm. At least, it was a paradox that did not deter his imagination from proceeding onward, beyond infinite sets.

Just as he had used finite sets to define stepping stones toward infinity, Cantor now used infinite sets to define stepping stones beyond infinity. His logic was exactly the same as before—the existence of an infinite set implies the existence of another, larger, set, which implies in turn the existence of another, still larger, set, and so on, ad infinitum. At each step, the larger infinite set is comprised of all the conceivable subsets of the infinite set before it. We can even state precisely the numerical leaps between Cantor's infinite stepping stones, using the very same formula as before. All it takes are some new symbols.

To denote the number of elements in an "ordinary" infinite set, Cantor invented the symbol $\aleph_0$ (pronounced "aleph null"), comprised of the first letter of the Hebrew alphabet and a subscript zero. According to our formula, then, an $\aleph_0$ set, such as the set of whole numbers, has precisely $2^{\aleph_0}$ conceivable subsets. This is a number—2 multiplied by itself $\aleph_0$ times—that is larger than $\aleph_0$ in the same way that 2^4 (sixteen) is larger than 4. It is the first stepping stone beyond infinity, the first transfinite number, which Cantor named $\aleph_1$ (pronounced aleph one). A set with $\aleph_1$ elements, in turn, has precisely $2^{\aleph_1}$ conceivable subsets. This is the second stepping stone, the second transfinite number $\aleph_2$, and so forth.

By any realizable standard, a transfinite number is fantastically, absurdly large. There are about ten billion (10,000,000,000) stars in the Milky Way, about sixty trillion (60,000,000,000,000) cells in the human body, and about three hundred quadrillion (300,000,000,000,000,000) seconds in the age of the universe. Furthermore, the estimated total number of protons in the universe, one of the largest numbers in nature, is written with a 1 followed by seventy-nine zeros. By comparison, if it were possible to write out explicitly the zeroth transfinite number, $\aleph_0$, it would be a 1 followed by an infinity of zeros. The transfinite number $\aleph_1$ would be a 1 followed by more than an infinity of zeros. Small wonder that the sequence of transfinite numbers ($\aleph_0$, $\aleph_1$, $\aleph_2$, $\aleph_3$, . . .) appeared to Cantor and others to be as boundless as the cosmos described by Immanuel Kant:

> It is natural to regard [the nebulous] stars as being . . . systems of many stars. [They] are just universes. It might further be conjectured that [all together] . . . they constitute again a still more immense system . . . which, perhaps, like

> the former, is yet again but one member in a new combination of members! We [on earth] see [but] the first members of a progressive relationship of worlds and systems; and the first part of this infinite progression enables us to recognize what must be conjectured of the whole. There is no end. . . .

In 1874, Cantor published the stepping-stone argument that had taken him from finite sets to infinite sets and beyond. At first, many mathematicians took him to task. Some of them dismissed his results summarily because they disagreed with the Platonic idea of treating infinity as though it were a noun. Others felt that Cantor had not pursued his argument to its logical conclusion.

To be consistent, they argued, he should treat the sequence of transfinite numbers as he had the sequence of whole numbers—as implying the existence of an actual aleph infinite ($\aleph_\infty$) set. With such a set, he could then recycle his argument to define an entirely new sequence of larger-than-$\aleph_\infty$, or trans-transfinite, sets. One might name them after the second Hebrew letter ב (beth), so that the new sequence would be $\beth_0$, $\beth_1$, $\beth_2$, and so on. Then, to continue being consistent, these mathematicians said, Cantor should treat this sequence, too, as though it implied the existence of a veritable beth infinite ($\beth_\infty$) set, and thus the entire logical argument could be repeated over and over and over again.

Although Cantor himself was never persuaded by this criticism, there are mathematicians today who do speak of an "absolute infinite." They denote it with the last letter of the Greek alphabet, omega (ω), and impute it to be the largest conceivable infinity. It is, by virtue of this definition, an infinity that we can never visualize; if we could, it would presumably be easy enough to visualize an infinity that was slightly larger. The mathematicians' ω is a label for something that we can never fully contemplate, and in this respect it is not too far different from the God described for us by St. Gregory:

"No matter how far our mind may have progressed in the contemplation of God, it does not attain to what he is, but to what is beneath Him."

Even without an ω, however, Cantor's set theory did find favor eventually with the mathematical community. David Hilbert, one of the most respected mathematicians at the turn of the century, hailed Cantor's work in 1910 as "the most admirable flower of the mathematical intellect and one of the highest achievements of purely rational human activity." Similarly, the English mathematician-philosopher Bertrand Russell lauded Cantor's accomplishments as "probably the greatest of which the age can boast."

One of the chief reasons for the praise then and now is that simply by using Cantor's notions of set and equivalence we are able to compare and to distinguish infinities that appeared to previous generations to be just one thing: inconceivably large. It seems as though by thinking in terms of set theory, we can extend the scale of our imagination, our perspective, interminably. And the results, for the most part, have been unexpected eye-openers.

The first surprise, discovered by Cantor, is that the infinite set of whole numbers is equivalent to the set of fractions. From a normal perspective this would not appear to be the case. Looking at the edge of a ruler, we see that between any two whole numbers there is room for an infinity of fractions; for this reason it would appear that there are infinitely more fractions than wholes. Appearances notwithstanding, Cantor proved that each and every conceivable fraction can be paired numerically with a whole number. Since there are $\aleph_0$ whole numbers, Cantor's proof meant that there are also $\aleph_0$ fractions.

Cantor also proved that the set of irrational numbers (decimals that cannot be rewritten as fractions or wholes) is larger than either the set of wholes or fractions. Once again, our conventional perspective makes it appeai that just the opposite might be true. Looking at the edge ot a ruler, we are at a loss to picture a way in which there could be any space among the infinite clutter of fractions to fit even a single irrational number, let alone more than one. And yet, according to Cantor's proof, between any two infinitely close fractions—fractions that are absolutely adjacent to one another—there dwells an infinity of irrational numbers, enough of them, in fact, that their total population exceeds the combined populations of fractions and wholes.

The proof itself, like the result, is considered remarkable by most mathematicians, if only because of its cleverness. Cantor began his argument by imagining that all the irrational numbers were listed (in no particular order) in a single column:

.17643567 . . .
.23482435 . . .
.62346286 . . .

and so on. The use of the ellipses is necessary because by definition each irrational number is an infinitely long string of random digits.

Cantor imagined next that each irrational number, each row of the interminable column, is paired with a whole number, beginning with 1:

1 .17643567 . . .
2 .23482435 . . .
3 .62346286 . . .

At this point in his argument, Cantor's purpose was to determine whether such a pairing would exhaust equally the wholes and the irrationals. If it did, then he would know that the irrationals, like the wholes, were $\aleph_0$ in number. Cantor proved, however, that it did not, by discovering that at least one irrational number will always be excluded from any such roster and therefore from being paired with a whole number.

There is a trick to finding that one irrational number.

Begin by taking the first digit of the first irrational number, followed by the second digit of the second irrational number, and so on. You will end up with the irrational number .133 . . . . Next, change each digit arbitrarily, so that you get, say, the irrational number .245 . . . . And this is it. This is an irrational number that is by design different from the first irrational in its first digit, different from the second irrational in its second digit, and so on down the roster. In short, it is an irrational number that is unlike any one of the irrationals that is paired with a whole number, which goes to prove that there are more irrational numbers than whole numbers.

It also proves that while you could, in principle, count your way one at a time through the infinity of whole numbers, you could never do the same with irrationals. There are more of them than wholes and therefore more of them than you could count. Cantor, recognizing this as a consequence of his proof, named the qualitative difference between the two infinities by referring thereafter to the wholes (and the fractions, by virtue of their being equal in number to the wholes) as a countable infinity and to the irrationals as a continuum or noncountable infinity.

Naming the qualitative difference, however, was easier for Cantor than determining the exact quantitative difference between the total populations of wholes and irrationals. To this day we still don't know exactly how many irrationals there are, although it has been established that the total number cannot be more than $\aleph_1$. Cantor himself guessed that the total number of irrationals is exactly $\aleph_1$, mainly since $\aleph_1$ is the next largest infinity after $\aleph_0$ defined by set theory. His guess came to be known as the Continuum Hypothesis. But there is still the uneliminated possibility that the number of irrationals actually lies somewhere *between* $\aleph_0$ and $\aleph_1$. Evidently, even by availing ourselves of set theory, our imagination is as yet not so incisive that we can make such fine distinctions and adjudge the Continuum Hypothesis to be true or false.

Besides using his notions of set and equivalence on arithmetical numbers, Cantor also used them on geometrical points, with results that surprised even him. He had intended simply to compare the populations of points in spaces of various dimensions, from the 1-D line, to the 2-D plane, to the 3-D volume, and so forth. And he had expected—or hoped—that his alephs would distinguish the various populations in some simple way, something on the order that in a space of x dimensions there are an $\aleph_x$ number of points. What he proved, however, was that there is an identical infinity of points in each and every space, irrespective of its dimension.

The points on a line, he showed, can be paired numerically with the points on a plane, which in turn can be paired numerically with the points in a volume, and so on. If geometrical points are thought of as clay, Cantor had shown that from the same-size lumps of clay you could fashion vessels of any number of different dimensions. The one caveat to his revelation is that the actual number of points being referred to here is just as uncertain as the number of irrationals.

This is not a coincidence. The number of points on a line is related directly to the number of irrationals, by way of the number line. A number line, which looks just like the edge of a ruler, is nothing more than a line whose points are paired with numbers: wholes, fractions and irrationals. Without knowing for certain how many irrationals there are, we cannot be sure of how many points there are altogether. Consequently, all we can say for now is that the number of points in a space of any dimension is always the same, always greater than the population of whole numbers, and always equivalent to the population of irrational numbers.

Cantor also showed that the population of points is the same regardless of the *size* of a space; there are as many points on this 2-D page as there are on all the pages of this book and in the whole 4-D universe. Had medieval philosophers been aware of Cantor's work, therefore, they would have known that you can accommodate as many point-size angels on a flat-headed pin, a tiny 2-D disk, as you can in all of heaven, a space of presumably infinite dimension and size.

Such imaginative and rational information about infinity as Cantor and other mathematicians before him have provided us not only puts us one up on Shaw's bushman, who can only count on his fingers, but it also inevitably upgrades our perceived relationship to the physical universe.

Not so long ago—that is, no more than a few thousand years—we had no numeral great enough to express the number of grains of sand on all the beaches on earth. Consequently, so far as we were concerned, it was as infinitely great as the distance to the stars. Then the mathematician Archimedes invented a numerical notation that enabled him to express gigantic quantities such as had never before been conceived. With his number system, which was a forerunner of our modern decimal system, Archimedes could calculate in tens of myriads, myriads of myriads, and myriads of myriads of myriads. In his treatise *The Sand Reckoner,* he actually calculated for the first time in human history the number of grains of sand on all the beaches in the world. Even more significantly, he went on to estimate that "the number of grains of sand which could be contained in a sphere the size of our 'universe' is less than 1000 units of the seventh order of numbers [which today we would write as 10^{52}, or a 1 with fifty-two zeros after it]."

That was most likely the first time anyone had used the words "less than" in describing the size of the universe. Certainly it was not the last time, for between then and now one mathematician after another has soared with the power of the human imagination to steal a view inaccessible to the mere human senses. And what they have seen has so dwarfed the physical universe that it is now proper to respond most affirmatively, I believe, to the rhetorical question put to us by Alexander Pope:

> He who through vast immensity can pierce,
> See worlds on worlds compose one universe,
> Observe how system into system runs,
> What other planets circle other suns,
> What vary'd being peoples every star,
> May tell why Heaven has made us as we are.
> But of this frame, the bearing and the ties,
> The strong connections, nice dependencies,
> Gradations just, has thy pervading soul
> Look'd thro? Or can a part contain the whole?

The physical universe no longer does contain us wholly, if it ever did. We are beings at once finite and infinite, in the sense that our physical selves are prisoners of a finite realm, but not so our imaginative selves. Ever since Cantor's reverie, part of us has been liberated from even the far limits set by Archimedes' myriads, and now we roam freely beyond the ordinary infinity of the ponderable universe.

Related Essays

Irrational Thinking
Singular Ideas
Nothing Like Common Sense
An Article of Faith

Natural Infinities

SINGULAR IDEAS

> To see the world in a grain of sand,
> And a heaven in a wild flower;
> Hold infinity in the palm of your hand,
> And eternity in an hour.
>
> William Blake, *Auguries of Innocence*

In mathematics, infinity is the name given to something that is bigger than our minds can imagine. For that reason, there have always been many mathematicians who oppose thinking of infinity as a definite something; to them, infinity names an incomplete, rather than a complete, concept. They prefer to think of infinity as an unending progression of definite things that we *are* able to imagine, such as the sequence of whole numbers 1, 2, 3, and so on.

For other mathematicians, infinity names a concept that is whole and definite. Among those who believed this—to the great advantage of mathematics—was the German Georg Cantor. About 100 years ago, Cantor developed a very influential theory of modern mathematics, called set theory, in which he assumed that infinity is something that can rationally be treated as a complete concept.

One question raised by this ancient and ongoing disagreement is whether scientists have any evidence for the existence in nature of a definite, fully contained infinity. If so, such evidence would provide mathematicians with at least a physical justification for positing the conceptual existence of an explicit infinity. And if not, perhaps those mathematicians are correct who argue that we have no basis for conceptualizing infinity as a complete entity.

As it happens, there is plenty of scientific evidence, old and new, to support the claim that infinity exists as a complete and definite concept. I am not speaking here of the apparent infinity of the cosmos, but of evidence of infinities that are sufficiently circumscribed or localized to be considered complete and definite things. The cosmos is evidence of just the opposite conclusion—that infinity is an unending progression, the whole of which we are unable to imagine, much less observe scientifically.

One of the oldest known incarnations of a localized infinity is the common electron. It was only in 1897 that an electron was actually first isolated in the laboratory, but as far back as 600 B.C., the Ionian mathematician-scientist Thales was speculating about tiny particles that exude an electric force. (He was one of the first to study the effects of electrostatic attraction created by rubbing a piece of amber with a wool cloth.) By the eighteenth century, we knew from the results of experiments with electrostatically charged spheres that the electric force of an electron gets weaker as we move away from the point-size particle. The most revealing experiments of this kind were conducted by the French engineer Charles Coulomb. In 1785, he found that if we increase our distance from the electron by two, the strength of the electric field diminishes by four.

The simple but startling implication of Coulomb's law, as it came to be called, is that the electric force gets stronger as we move toward the electron; in particular, if we halve the distance between us and the electron, the strength of its electric force quadruples. If we are actually upon the electron itself, the electric force is infinitely strong, the embodiment of a mathematical infinity. And because the infinity is not infinitely expansive, but rather confined to a single point, the electron is indeed a localized infinity—and a portable one at that, since electrons are free to move around.

Unfortunately, since our eyes are not designed to enable us to actually see these corporeal infinities, we must settle for seeing them through the electronic senses of scientific instruments such as Geiger counters and spark chambers. (These instruments merely chirp or spark in the presence of an electron, however, and I find it hard to believe that such modest exclamations are fair indicators of what localized infinities must really look like!)

There *is* another significant example in nature of a localized infinity, however, and this one we might actually be able to see. It is a black hole, the burned-out, superdense remains of a once-active star.

When a star consumes all its fuel (which is mostly hydrogen), its matter cools and darkens. Since there is no longer any outward pressure from combustion to counterbalance the inward pressure from gravity, the star collapses under its own weight. If the star is not too massive to begin with (that is, with a mass no more than about three times that of our sun), it will collapse only so far and then stop. It will usually end up being from ten miles to five thousand miles across, which means that its mass ends up being concentrated into a relatively small volume (by comparison, the sun is presently about eight hundred thousand miles across). The mass densities for such dead stars, the so-called white dwarfs and neutron stars, are therefore quite great; a chunk the size of a sugar cube can weigh as much as 100 tons.

Mind you, that's what happens if the star is not very massive to start with. If it is, then the dying star will collapse all the way down to a point—literally—when it dies. In such a case, the resulting density is infinite, because all the mass is compressed into a volume that is really nil.

This unimaginably dense object, this localized infinity, is called a black hole because its gravitational force is so powerful that even light rays that wander into its vicinity get pulled in. If an object (such as a spent satellite) gets pulled toward the earth, that object can in principle escape with enough power; escape from a black hole, however, is impossible once an object has begun to fall in—that is, it would require an infinite amount of power for an object to pull free of a black hole.

Precisely because a black hole does behave like an irresistible vacuum cleaner in the region immediately around it, it had been assumed until recently that a black hole seen from a distance would look like just that—a black hole; we would have no chance of seeing the point of infinite density itself, since it dwells at the heart of the opaque region. Any light the localized infinity might be giving off, it was assumed, would be unable to escape and reach our eyes.

During the last several years, however, some astronomers have theorized that it is possible that if a black hole is spinning fast enough, then it will shed its opacity. The localized infinity within it would thereby be exposed, they say, and there is even the possibility that it would spew out great quantities of radiation, much like some cosmic Fourth-of-July pinwheel.

This would be a rare chance for us to actually see a localized infinity, but there is still some question about this theory. So far, only one physicist (Joseph Weber at the University of Maryland), claims to have observed radiation coming from the center of our galaxy such as that which we might expect from a spinning black hole. However, other scientists who have looked for this radiation source have not been able to find it. For now, matters remain ambiguous with respect to spinning black holes.

With regard to "ordinary" black holes, however, most astronomers are now convinced that they have located several of them out among the stars. Astronomers are especially certain that one dwells in the midst of the constellation Cygnus, in tandem with an ordinary star. The reason for their certainty is that through studying the erratic movements of the ordinary star, they were able to infer the presence of a nearby massive object—yet when they looked for this neighboring object they could not find it. Their conclusion is that this invisible companion of the ordinary star is a black hole, and they have named it Cygnus X-1.

Ordinary black holes have also been recently predicted, and perhaps have already been seen, to come in sizes smaller than originally expected. According to modern astronomy, these localized infinities shrouded in darkness can get to be as small as atomic nuclei—about .0000000000001 centimeters across—and are free to move around like electrons. These mini–black holes are thought to have been formed ten billion years ago by the enormous pressures and temperatures believed to have been in existence when the universe was in its infancy. Although they are tiny, they are nonetheless quite massive (typically, they have the mass of an iceberg) and so could conceivably cause a great deal of damage to anything they might run into.

Ever since the idea of mini–black holes was discussed by astronomers, in fact, many people have ascribed the mysterious 1908 explosion in Tunguska, in central Siberia, to the collision of one of these ancient mariners with the earth. Residents of Tunguska at the time reported seeing a fireball crossing the sky that was so bright that "it made even the light from the sun seem dark." There's really no way of being certain about what actually happened at Tunguska, of course, but in principle a mini–black hole collision would give off a great deal of energy—enough, by some calculations, to equal the power of a small atomic bomb.

Localized infinities such as the electron and the black hole are what scientists call singularities. They are, as we have seen, points in space (and also in time, in some cases) where some physical quantity is infinitely large. As such, their existence mitigates the argument that there is no rational basis for a discussion of infinity as if it were something whole and definite. A singularity is an infinity that you could, in principle, hold in the palm of your hand—in fact, in the case of electrons, trillions and trillions actually do exist in the skin of your palm.

Since they exist physically, it does not seem to be unwarranted to defend their existence conceptually through mathematics. This is not to say that in recognizing the existence of physical singularities we are any nearer to being able to imagine the largeness of infinity; but we are, as a result, better able to justify imagining how such largeness can be thought of as a whole entity with definite boundaries, rather than only as an ever-expanding frontier without well-defined boundaries.

Also, even if scientists had no evidence for singularities such as electrons and black holes, there would still be one major piece of evidence in favor of referring to infinity as a whole and definite thing, evidence that has been around since we first conceived of infinity, named it, and began debating its fine qualities. The oldest piece of evidence for a fully contained infinity is the mind itself. For, although infinity is confined in the mind within a volume that is relatively large compared to that of an electron or a black hole, the mind is no less singular than these.

Related Essays

Beyond Infinity

Applying Abstract Mathematics
INVENTING REALITY

That minister of ministers,
    Imagination, gathers up
The undiscovered Universe,
    Like jewels in a jasper cup.

John Davidson, *There Is a Dish to Hold the Sea*

The Greek philosopher Plato believed that anything and everything conceivable exists somewhere in the universe. Many learned Europeans of the eighteenth century, including the English philosopher John Locke, also held this view, and it led them to believe reported sightings of mermaids and other oddities of the human imagination. This notion that if something can exist then it must exist is known as the principle of plenitude.

Today it remains a dubious hypothesis where the objects of our general imagination are concerned, but it is a demonstrable verity where the objects of the mathematical imagination are concerned—and many of these are every bit as extraordinary as any mermaid might be. The principle has been vindicated in every branch of mathematics, but nowhere more clearly so than in algebra (the study of arithmetical relationships among numbers). Each time during the past four centuries that algebraists have encountered a new, bizarre kind of number, scientists have discovered that it could be applied to describe something real. This evidence in favor of the principle of plenitude is all the more noteworthy because, as often as not, ideas in mathematics are conceived by playful, imaginative minds whose first concern is to be rational, not realistic—so the extensive coincidence between the mathematicians' invented world and the natural world is not merely the result of a purpose on the part of mathematicians to describe reality.

In algebra, the mathematician's purpose is to study all the ways numbers are related through the standard arithmetical operations of addition, subtraction, multiplication, and division. For instance, in the study of simple additive relationships such as $3 + 4 = 7$, algebraists have proved that a whole number added to a whole number will always equal a whole number. Their way of stating this is to say that the whole numbers are "closed" under addition; their shorthand way is to write $x + y = z$, in which the letters represent whole numbers.

Algebraists use letters as stand-ins for numbers because it enables them to express concisely an arithmetical relationship that is generic to an entire class of numbers. When all but one of the letters in a generic relationship is substituted with a specific number, as in $x + 4 = 6$, the remaining letter x takes on the role of an unknown number. Here it stands for the one and only whole number that will satisfy the equation, which for the algebraist poses a problem to be solved.

The equation $x + 4 = 6$ is one of the simplest algebraic equations imaginable; most of us could probably solve it in our heads (the solution is $x = 2$). More complex algebraic equations cannot be solved so easily (for example, in the problem of finding a whole number x that satisfies the equation $x^3 + x^2 + 3x + 4 = 96$; the answer: $x = 4$), and over the years algebraists have spent a good deal of time figuring out ways of solving them.

Back in the sixteenth century, algebraists acknowledged the existence of only positive numbers, though they had long since learned from solving certain algebraic equations that x sometimes turned out to stand for other kinds of numbers. In the equation $x + 3 = 2$, for instance, x stands for -1. But minus numbers—or negative numbers, as they are now called—were an enigma to algebraists 400 years ago. Positive numbers could be conceptualized in terms of tangibles such as pebbles or marks on a sheet of paper, but it was difficult to accept the mathematical existence of something that represented "less than nothing," as the seventeenth-century French philosopher-mathematician René Descartes complained.

Without a logical way of thinking about negative numbers, without some conceptual model, algebraists were unable to comprehend what it meant to add, subtract, multiply, and divide negative numbers. For that reason, negative numbers were not perceived as legitimate objects of algebraic study; their presence in certain algebraic equations was taken to have no greater significance than the existence of nonsense words in a language.

Not until late in the eighteenth century did algebraists learn how standard arithmetical operations applied to negative numbers, which became associated with the concept of debt. For instance, if there is a negative \$10 in your checking account, this means that your account is overdrawn by that amount. Subtracting a negative \$10 is the same as adding a positive \$10, because, as noted by the Swiss mathematician Leonhard Euler in 1770, "to cancel a debt signifies the same as giving a gift."

Algebraists also articulated the rule for multiplying negative numbers, something that is easy to illustrate in terms of an ordinary situation. Consider, for instance, that there are effectively two kinds of voters in any election: positive and negative. A positive voter is someone who actually casts a ballot and thereby exercises a positive influence on the election. A negative voter is someone who is qualified to vote but fails to cast a ballot; in a negative, or indirect, way, such a voter also exercises an influence on the election's outcome.

To see what this example has to do with the rule for multiplying negative numbers, imagine that a ballot actually cast for some proposition is worth a positive ten points and that a ballot cast against it is worth a negative ten points. There are effectively two ways that the proposition can gain an advantage in the outcome: either by the actual casting of a pro ballot by a positive voter or by the failure of a negative voter to cast a con ballot.

The first way illustrates the ordinary multiplication of positive numbers. That is, if five positive voters who are for the proposition ($+5$) actually cast their pro ballots (at $+10$ points each), then the proposition gains a total of fifty (5×10) positive points.

The second way illustrates the ordinary multiplication of negative numbers. If five negative voters who are against the proposition (-5) all neglect to cast their con ballots (at -10 points each), then in this way too the proposition has gained, indirectly, a total of fifty (-5×-10) positive points. It is a fifty-point advantage that the proposition would otherwise not have had.

In short, as with a positive number times a positive number, the result of a negative number times a negative number is also always a positive number. This rule is often called the law of signs, and of all the algebraic rules for combining negative numbers it seems to perplex people the most. W. H. Auden, for one, expressed his impatience with the law of signs in just two lines:

> Minus times minus is plus,
> The reason for this we need not discuss.

However, the law of signs is really no more mysterious than

the double negative in language, something with which most of us are not unfamiliar.

In 1930, the English scientist P. A. M. Dirac applied the concept of negative numbers to his theoretical studies of nuclear physics and came up with the idea of negative matter, or antimatter, as it is now called. Specifically, Dirac predicted the existence of a new elementary particle that he likened to a negative electron. According to his theory, this particle would have the mass of an electron but would have a positive rather than a negative electric charge, and if it were ever in the vicinity of an electron, both particles would be annihilated instantly.

This last part was the most unusual feature of the positron, as Dirac named his theoretical particle, yet it was his best guess of what would happen to an ordinary particle such as an electron if it were to be combined with its negative counterpart—as with the addition of a negative 10 to a positive 10, the result was apt to be zero. In 1932, Dirac's prediction was confirmed in every respect when the American physicist C. D. Anderson identified the tracks of positrons in his cloud chamber.

Since then, a myriad of laboratory experiments have confirmed the existence of positrons and have turned up negative counterparts to all other known subnuclear particles, including the proton and neutron. As a consequence, antimatter is now recognized to be as significant a part of the natural world as negative numbers are of the algebraic realm. It exists in abundance at the subnuclear level, though for reasons not well understood, there is a notable absence of it in our earthly environment. Some astronomers speculate, however, that stars and galaxies made entirely of antimatter may be strewn throughout the universe.

Today, though scientific theories agree that antimatter is the embodiment of mathematical negativity, they differ in their explanations of what exactly about it is negative.

According to Dirac's theory, the explanation has something to do with energy. As Dirac puts it, the presence of a positron with positive energy is the physical manifestation of the absence of an electron with negative energy. Positrons, in other words, are as positively sensible as the absence of any sound is positively silent.

According to the American Nobel Prize–winning physicist Richard Feynman, the explanation of antimatter's negativity has something to do with time. Specifically, Feynman says that a positron moving forward in time is the physical equivalent of an electron moving backward in time. Here, forward is meant to be associated with the positive direction and backward with the opposite, negative direction.

Although there are certain technical advantages to Feynman's idea, scientists do not yet have any empirical evidence with which to make a final choice between the two theories. Nonetheless, because both theories use the concept of negative quantities, it is accurate to say that particles of antimatter are the mathematicians' negative numbers incarnate.

If negative numbers were to sixteenth-century algebra what Kandinsky's haystacks were to nineteenth-century painting, then the so-called imaginary numbers were to sixteenth-century algebra what Picasso's cubism was to twentieth-century painting. Imaginary numbers are numbers that are neither positive nor negative. And yet, like negative numbers, they forced themselves upon algebraists by cropping up here and there in solutions of certain algebraic equations.

Take an ordinary equation such as "x times x equals 4" ($x \cdot x = 4$) or, equivalently, "x squared equals 4" ($x^2 = 4$). This equation is satisfied when x is the square root of 4, or 2. Two is a number which when multiplied by itself is 4. In the algebraists' jargon, 4 is the square of 2 ($4 = 2^2$) and, conversely, 2 is the square root of 4 ($2 = \sqrt{4}$).

In "x squared equals negative 1" ($x^2 = -1$), the equation is satisfied when x is the square root of negative 1 ($\sqrt{-1}$), but that happens to be neither a positive nor a negative number. It is not positive, because a positive number squared is always a positive number. And it is not negative, because (per the law of signs) a negative number squared is also always a positive number, and the x in the equation is a number which when squared is a negative number (-1). Therefore, the x in question—the square root of negative 1—is neither a positive nor a negative number.

Realizing this, algebraists of the sixteenth and seventeenth centuries were hard put to know what to think of the number $\sqrt{-1}$ except that it was probably an irrational, inconsequential fluke. Descartes called $\sqrt{-1}$ and other square roots of negative numbers "imaginary," and the German mathematician Gottfried Leibnitz referred to them as "a wonderful flight of God's Spirit."

These enigmatic appellations were the only descriptions that algebraists at the time could manage, because they did not have any logical way of thinking about imaginary numbers. And without some conceptual model for imaginary numbers, algebraists were unable to comprehend what it meant to add, subtract, divide, and multiply these ineffable quantities. From that perspective, there was no reason to believe that these imaginary numbers were a sensible part of algebra.

It was not until 1797, when the Norwegian surveyor Casper Wessel discovered a way to conceptualize them, that imaginary numbers were finally recognized by algebraists to have a proper place alongside the real numbers (as the positive and negative numbers together had come to be called). The key was in thinking of imaginary numbers as existing mathematically in a different dimension than that of real numbers—in other words, if the real numbers were imagined to be the coordinates of longitude, then the imaginary numbers would be the coordinates of latitude.

With this simple model in mind, Wessel was able to see just how to extend the meaning of addition, subtraction, multiplication, and division to include the imaginary numbers in algebra. For instance, combining a real number with an imaginary number is not like adding two real numbers. It is similar to combining a longitudinal coordinate with a latitudinal coordinate: the result is not a numerical sum, but instead is like the coordinates of a point on a surface.

Each point on the earth's surface or on a map is codified in terms of its longitudinal and latitudinal coordinates. Thus, according to Wessel's conceptualization, the algebraic relationships between real and imaginary numbers can be understood entirely in terms of a hypothetical maplike surface whose every point is labeled by a pair of numbers, with the imaginary number being measured along a latitudinal (or vertical) scale and the real number measured along a longitudinal (or horizontal) scale. Wessel named this mathematical surface the complex plane.

Some hundred years following Wessel's revelation, Albert Einstein and his contemporaries applied the concept of imaginary numbers to their studies of space and time. One of the results was the special theory of relativity, in which time and space are posited to be two different physical dimensions that are like the latitude and longitude of the universe. This means that a point in the universe is fully specified only if both its spatial and temporal locations are given. This part of the theory, at least, jibes with common experience, for whenever we're arranging to meet with someone we're always certain to specify a place *and* time. According to this theory, the only accurate map of the universe is one in which the spatial (or latitudinal) dimension is measured along a scale of real numbers and the temporal (or longitudinal) dimension is measured along a scale of imaginary numbers.

This would mean that the universe is a physical manifestation of the algebraists' complex place, and every experiment done to test the theory bears this out. Since 1905, when Einstein first articulated the special theory of relativity, laboratory measurements have verified that the universe is a place in which moving clocks tick more slowly than stationary clocks, and objects seem to shrink along the direction in which they move. These, according to Einstein, are the telltale traits of a universe that is the embodiment of a mathematical complex surface.

With imaginary numbers included, traditional algebra—that is, as it has been studied since Egyptian times—is a self-contained subject: The positive, negative, and imaginary numbers together constitute a single algebraic species. We may call them the traditional numbers. If you add, subtract, multiply, and divide two or more traditional numbers, the result will always be a traditional number.

This does not mean, however, that algebra stagnated as a subject after 1797, when Wessel made his discovery concerning imaginary numbers. Since the early 1800s, algebraists have invented entirely new algebraic species to play with, and the result has been a plethora of abstract algebras, each self-contained. In each abstract algebra, the numbers of traditional algebra have been replaced with abstract numbers whose only resemblance to traditional numbers is that in some logical sense they, too, can be added, subtracted, divided, and multiplied in a self-contained manner. Imagine for a moment that positive, negative, and imaginary numbers are to algebraic species what the different human races are to animal species. Then the various species of abstract numbers are as different from traditional numbers, and from each other, as elephants or giraffes or wallabies are from humans.

Also, since the 1840s, algebraists have gone beyond imagining abstract algebraic species to imagine "super-abstract" algebraic species, as I will refer to them here. With traditional and abstract numbers, it doesn't matter in which order two numbers are multiplied; the answer is the same either way. For example, $2 \times 3 = 6$, and so does 3×2. Numbers obeying this algebraic property are called commutative. Traditional and abstract numbers are commutative, but super-abstract numbers are not; by not obeying a property of numbers that most of us take for granted, they are that much more removed from our common experience with numbers.

As with the traditional numbers, so many of these abstract and super-abstract algebraic species have been applied successfully to describe reality that the principle of plenitude is well on its way to becoming a fait accompli where the inventions of the algebraic mind are concerned.

For example, one super-abstract species of numbers called matrices has proven to be precisely what scientists needed to describe subatomic physics. Matrices were invented contemporaneously, in 1860, by the British mathematicians James J. Sylvester and Arthur Cayley. Some sixty years later, the German physicist Werner Heisenberg, having recently learned about matrices, used them to articulate an entire theory of how things behave at subatomic scales.

Heisenberg's theory came to be known as the theory of quantum mechanics, and in the decades since its development, numerous tests of its predictions have convinced most scientists of its credibility. It is a theory replete with revolutionary theses about the subatomic realm, each one related to some algebraic property of matrices. For example, the noncommutativity of matrices is related to the thesis that when an intrusion caused by the act of measuring is significant enough to alter whatever is being measured, then it matters in which order two or more measurements are made: If two or more matrices are multiplied, the results will be different depending on the order in which they're multiplied. It's like saying that if you taste the wine in your glass first and weigh it second, then you will get a different result than if you reverse the two measurements.

Before quantum mechanics, scientists had taken it for granted that if only they were discreet enough with their measurements, it would be possible for them to observe the natural world without disturbing it. And, indeed, this is true for many domains of the natural world. In the domain of our common experience, for example, it really would not matter if a doctor listened to our heartbeat first and took our temperature second or vice versa. In describing commutative measurements of this sort, scientists use traditional numbers.

According to the theory of quantum mechanics, however, it is in principle not possible for scientists making measurements on the subatomic world to ever eliminate entirely the intrusion their observations cause. This thesis is known as the Heisenberg uncertainty principle and it is a physical manifestation of the noncommutativity of matrices. In other words, our efforts to poke about the tiny watchworks of the natural world will always be fat-fingered.

The evidence in algebra for the principle of plenitude, combined with similar evidence in other branches of mathematics, makes it seem as if the mathematical imagination is a sixth sense. If the physical equivalents of only a few isolated mathematical ideas had been found to date, then the coincidence might be nothing more than chance. But, in fact, the coincidence is great enough to have compelled the German mathematician, physicist, and Nobel laureate Eugene Wigner to speak of it as the "unreasonable effectiveness of mathematics" in describing reality.

It doesn't appear that we are merely inventing ideas that just happen to describe sensible objects; it seems, rather, that the mathematical imagination is an extra sense with which we can perceive the natural world. And it is an extremely efficacious sense, because it often perceives reality long before our scientific senses do. If thought of in this way, the coincidence between the natural world and the mathematical world is not any more mysterious than the coincidences between the natural world and the auditory, tactile, and olfactory worlds. The coincidence is evidence for the common thesis, in that case, that our senses corroborate one another simply because they all perceive different aspects of a single reality.

Thinking of the mathematical imagination as a sixth sense leads me naturally to wonder whether the human imagination generally ought to be thought of as a sense. If it is, then we can expect that mermaids, UFOs, ghosts, and all the other creations of the human imagination will one day be discovered; then again, perhaps these are merely the aberrations of a sixth sense that can occasionally deceive us into believing that we see something that is not actually there. If the latter is true, perhaps there is something about the mathematical imagination in particular—its rationality, for example—that makes it less susceptible to conceiving of illusory things. That would account for the fact that inventions of mathematics have such a high rate of being verified as real.

If this mathematical sixth sense does exist, then certainly modern biologists would have us know that it probably evolved, like all the animal senses, in order to enhance our survivability as a species. "Perceiving . . . reality," as the French biologist and Nobel laureate François Jacob put it, "is a biological necessity." If the mathematical imagination is our sixth sense, therefore, the traditional, abstract, and superabstract numbers of algebra are not merely algebraists' playthings, they are, by dint of helping us to perceive the world around us more accurately, instruments of our survival.

Related Essay

Abstract Symmetry

Group Theory

ABSTRACT SYMMETRY

Tiger, Tiger, burning bright
In the forests of the night,
What immortal hand or eye
Could frame thy fearful symmetry?

William Blake, *Songs of Experience*

The architecture of the cities of the earth would reveal us to any alien as being essentially visual creatures. Buildings are designed with arched entranceways and mirrored, stained, or louvered windows—details that suggest the importance of a building's appearance to its architects. The alien, in recognizing the dominating theme of symmetry in the shapes of our constructions, might also surmise that we are esthetically sophisticated beings.

The situation is the same in *our* observations of *nature's* architecture. We are like that alien as we study the earth and universe around us with curious senses. At first, we relied solely on our five senses, and in that way we were sensitive to the ostensible symmetries that so characterize the natural world, from the visual symmetry of the starfish to the aural symmetry of a bird's song. In the past 150 years, however, we have acquired a sense for abstract symmetries that can be perceived by the mind alone. The sense is provided to us by group theory, the mathematical study of symmetry, and with it, nature now looks even more symmetrical, more esthetic, than ever before.

The invention of group theory is generally credited to the singularly tragic figure of Évariste Galois, a short-tempered Frenchman killed at the age of twenty in a senseless duel over, in his words, "an infamous coquette." In 1829, at the age of seventeen, he wrote his first paper on computing the roots of fifth-degree algebraic equations, a subject that would lead him directly to group theory.

Algebra is the mathematical study of numerical relationships, and algebraic equations are the means used to express those relationships. If we think of equations as being to algebra what sentences are to English, then the roots of an equation correspond to adjectives and a fifth-degree algebraic equation is like a sentence with five adjectives.

Algebraic equations come in all degrees and are cataloged that way by mathematicians. Just as the number of adjectives in a sentence reflects the complexity of the thought being expressed, the degree of an algebraic equation reflects the complexity of the numerical relationship it expresses. The numerical relationships that concerned Galois were of such complexity that each one required exactly five roots to fully describe it.

In his first paper, Galois was specifically seeking a systematic way of finding those roots in order to eliminate the haphazard procedures that algebraists had theretofore had to resort to—it was as though they had been having to come up with all of the adjectives without the aid of any dictionary or thesaurus.

Past efforts had led to the discovery of systematic procedures for discerning the roots of lesser-degree equations: A procedure for second-degree equations had been known to the Babylonians, for instance, and methods for third- and fourth-degree equations had been developed in the 1500s by the Italian mathematicians Scipione dal Ferro, Niccolo Fontana, and Lodovico Ferrari. Since that time, however, no one had succeeded in finding an orderly mathematical way to determine the roots of fifth- and higher-degree equations.

Galois's first paper was a noble but unsuccessful effort. Then, two years later, he made two historically significant discoveries: He found that no systematic procedure exists for finding the roots of fifth- and higher-degree equations and that some such equations have numbers as roots that do not normally fall within the realm of algebra. Today these numbers are called the transcendental numbers, and in our analogy between algebra and English they would correspond to adjectives in foreign languages.

The second part of Galois's discovery was akin to our saying that when our thoughts reach a certain level of complexity, some of them will defy being described properly with the adjectives of any single language. For some people—multilingual poets, for example—such a result is liable to sound like an indisputable truism. For the rest of us, who often struggle unsuccessfully to think of *le mot juste,* it probably sounds at least plausible.

It is, in any case, the mathematical particulars of this part of his discovery that entitles Galois to be remembered as the founder of group theory. The roots of fifth-degree equations behave differently when interchanged, he noticed, depending on whether or not they are all algebraic numbers.

If they are, then interchanging them in the equation will result in another equation that is as algebraically sensible as the first one. This is like the sentence, "She is a robust, flamboyant, wealthy, extravagant, gregarious woman"; rearrange the adjectives, and the sentence still sounds perfectly sensible. On the other hand, if some of the roots are transcendental numbers, interchanging them results in an algebraically nonsensical equation. Take the sentence, "He is a robust, flamboyant, wealthy, extravagant, Belgian man"; if "Belgian" were given the position of any of the other adjectives, the result would be an incorrect sentence.

Although Galois himself did not live long enough to explicate all the implications of this particular discovery, we now realize that it is associated with the concept of symmetry. This association, which is a cardinal feature of group theory, is abstract; it can be illustrated by considering the five-pointed spatial symmetry of a starfish.

Imagine that the starfish is oriented so that one of its arms points straight up. A mathematician would have you notice that if you rotate the starfish through zero, 72, 144, 216, or 288 degrees, it will appear just as it did to begin with, having one of its arms pointing straight up. This property of the starfish, which is a direct consequence of its symmetry, is what group theorists refer to as "invariance under a group of rotations."

If we now imagine that the starfish's arms are numbered one through five, then it becomes clear that rotating the starfish through one of its invariant angles is represented abstractly by interchanging the numbers labeling its arms. Rotating the starfish through, say, 72 degrees is represented by the numerical interchanges: One goes to two, two goes to three, three goes to four, four goes to five, and five goes to one. With this abstract way of representing things, the starfish's symmetry is described not by its invariance under a group of rotations but by its invariance under a group of numerical interchanges.

It is in this abstract sense that group theory associates interchanging numbers with symmetry, and that Galois's revelation about algebraic roots is tantamount to perceiving that numerical relationships can be just as symmetrical as starfish.

According to group theory, those relationships are precisely the ones described by fifth-degree algebraic equations whose roots are entirely algebraic numbers (that is, with no transcendental numbers). In an abstract sense, the theory explains, such a numerical relationship has the symmetry of a five-pointed starfish, because interchanging the five roots does not alter the algebraic sensibility of the equation that describes the relationship. In short, it is symmetrical like the starfish because, in its own way, it is invariant under numerical interchanges just as the starfish is.

By contrast, those numerical relationships that are described by fifth-degree algebraic equations whose roots include some transcendental numbers are not symmetrical. They are like a starfish whose arms are arranged randomly on its body. If such a starfish is initially oriented with one arm straight up, then no matter how we rotate it, it can never be made to look the same—it can only have the same orientation that it did originally if it is given one complete rotation. Along with its symmetry, the starfish has lost its invariance under a group of rotations, or equivalently, under a group of numerical interchanges.

For the same reason, according to group theory, numerical relationships described by roots that include transcendental numbers are as asymmetrical as our hypothetically misshapen starfish. The only difference is that the asymmetry of the latter is readily visible, while the asymmetry of the former, though no less real, can be perceived only through an understanding of group theory.

This makes group theory a true extra sense, and with it scientists have come to learn just how much of the esthetic symmetry of nature they were insensitive to when they had the use of only their five senses. Group theory has enabled them to recognize the symmetry of things merely by looking at the mathematical equations that describe them. Generally speaking, if something about an equation remains invariant under a group of numerical interchanges, it indicates that whatever the equation describes is symmetrical, be it physically or abstractly so.

Physicists use group theory to help them picture the various symmetrical shapes their theories impute to atoms.

According to these theories, atoms are nuclei enveloped in electronic fogs shaped like spheres, concentric spheres, dumbbells, yo-yos, and other symmetrical objects.

Similarly, chemists use group theory to help them visualize symmetry at the molecular level. They have discovered, for instance, that the shape alone of a molecule will affect its physical properties. Molecules with the symmetry of a cube taste and smell different from those with the symmetry of a pyramid.

Of the myriad extrasensory insights provided by group theory, the most remarkable is probably the one that has revealed to us a certain symmetry of the universe itself. Although it is a symmetry of space and time, two readily observable quantities, it is not observable through the senses, and would not be even if we were somehow able to pull away from the universe for a bird's-eye view. According to group theory, it is a symmetry that is associated with the law of conservation of energy and momentum. (Momentum is a physical quantity related to the motion of objects.) This law, which is believed by physicists to reign universally, states that the total quantity of energy and momentum in the universe is neither diminished nor enhanced by all the changes the universe sustains.

In group theoretical terms, the law of energy and momentum conservation is an invariance property of a symmetrical object, the universe. If we were to use an equation to tally up the total amount of energy and momentum in the universe at any given moment, we would find that the amount does not change from one moment to another. In other words, the universe is such that nothing is changed by interchanging the value of the equation at one moment with the value at another moment. According to group theory, this invariance under numerical interchanges means that the universe is symmetrical in some abstract sense.

An ability to even contemplate such matters as cosmic symmetries inevitably has the effect of sensitizing us to the beauty of the workings of natural laws, even when those workings may not seem esthetically pleasing to our eyes. Thus, group theory serves us as artists in an alien world, not merely as observers of one. We find pleasure in the symmetries of nature just as we do in the symmetries of our own architecture. How less pleasing might the universe appear to us were we, in Hesse's words, "poets without verse, painters without brush, musicians without sound," and, might I add, mathematicians without group theory.

Related Essay

Inventing Reality

Dimension

A REALM OF MANIFOLD POSSIBILITIES

A horizon is nothing save the limit of our sight.

Rossiter Raymond, *A Commendatory Prayer*

"Oh, yes, but he's such a one-dimensional person. . . ."

I overheard that unflattering dismissal of a successful businessman at a recent cocktail party and was reminded of Edwin A. Abbott's fictional world, Flatland. Even in that mythical two-dimensional land, being one-dimensional is disdainful: One-dimensional creatures, the straight lines, are the untouchables. Everyone else is a two-dimensional polygon of some sort.

I myself would have hesitated to restrict that businessman to one dimension. Experience suggests that the better you come to know anyone or anything, the more complex he, she, or it becomes—in other words, the more dimensions emerge. There is other evidence that dimension is in the eye of the beholder. Psychologists have learned that infants crawling on a glass floor will not hesitate to crawl past the edge of a steep cliff. They are not afraid of heights because they apparently do not *perceive* height; theirs is a strictly two-dimensional world, and only when they have developed further are they able to perceive the world more correctly as three- and four-dimensional.

In the course of human history, too, our developing perception of the natural world has imputed an increasing number of dimensions to the universe. For most of that time, our perceptual development has lagged behind the development of mathematical thinking on the subject of dimensions. The latter has evolved to such an extent that mathematicians today speak without flinching of infinitely dimensioned worlds and of objects with a fractional number of dimensions. Since much of this evolution has occurred only in the last hundred years or so, we might well wonder how expansive the universe will appear to us—dimensionally and otherwise—a hundred years from now.

Two thousand years ago, the Greeks perceived the universe to be three-dimensional, a perception supported by the senses and by the tenets of Euclid's geometry. All around them, the Greeks saw what we see today: a world filled with objects having length, width, and depth. It was natural for them to imagine that the receptacle that contained these objects itself had length, width, and depth.

According to Euclid, these qualities of length, width, and depth corresponded with what he meant mathematically by the term *dimension*. In Euclidean geometry, a line is imputed with the sole quality of length and is thereby defined as the quintessential one-dimensional object. A plane, which is given the qualities of length and width, is the archetypal two-dimensional object. And a solid, which is endowed with all three qualities of length, width, and depth, is defined as the model of a three-dimensional object. Thus, the mathematics of Euclid's time teamed up with common sense to support the ancient Greeks' impression of a 3-D universe.

For generations following Euclid, the universe continued to be seen as three-dimensional. Any speculation about a fourth dimension was usually dismissed as mathematically inconceivable. At one point, even the great Alexandrian astronomer Ptolemy discredited the idea of a fourth dimension. He did this by illustrating that although it was possible to draw three mutually perpendicular axes in space, it was not feasible to draw a fourth such axis.

There were, inevitably, those who speculated about the existence of a fourth dimension anyway. But until there was a mathematical or logical basis for them, such arguments tended to be mystical. One English philosopher of the mid-seventeenth century, Henry More, went so far as to insist that ghosts existed and were the inhabitants of the fourth dimension. Arguments such as these lacked scientific credibility, however, so they did not have a permanent influence on the continuing, widespread impression of a Euclidean 3-D universe.

The prejudice of this impression was still so gripping in More's time that when the French mathematician René Descartes expanded the language of Euclidean geometry in a way that introduced the mathematical possibility of a fourth dimension, he automatically discarded the possibility as unrealistic.

Descartes's approach to geometry differed from Euclid's in that he defined the dimensionality of the line, plane, and solid in terms other than length, width, and depth. According to his theory of analytic geometry, the dimension of an object is correlated with the number of coordinates that is required to map it. For example, a line is one-dimensional because it can be mapped using only one coordinate. Imagine the line to be a street; every point on the line, or every house on the street, can be uniquely labeled with a single numeral. A plane, which we might compare to a flat earth, is two-dimensional, because two coordinates are needed to map it—each point on the plane, like each location on earth, is uniquely labeled with two numerals, akin to longitude and latitude. Similarly, a solid is three-dimensional, because three coordinates must be used to map it.

Descartes's definition of dimension was significant at the time not because it was better than Euclid's, but because it was quantitative rather than qualitative and was based more on our logical capacities than on our sensory experiences. Whereas Euclid relied on our understanding of the qualities of shape (length, width, and depth), Descartes relied on our understanding of the logic of an analytic process (a process resembling multidimensioned cartography).

If Descartes's theory of analytic geometry had come out at a time less persuaded by sensory experience and Euclidean thinking, chances are that mathematicians would have unanimously recognized the logic of a four-dimensional object. It would only have required of them the recognition that such an object is the mathematical entity that requires four coordinates to map it.

Logical though this inference was, it was not compelling enough to overcome the reluctance of mathematicians to acknowledge even the possibility of something that they could not visualize. Arguments such as Ptolemy's continued to hold sway, and consequently, analytic geometry's implication of a mathematical fourth dimension was lost not just on Descartes himself and on his generation but also on several more generations thereafter.

It was not until 1854, when the young German mathematician Bernhard Riemann announced a further extension of Euclid's geometry and Descartes's analytic geometry, that the idea of a fourth dimension was mathematically recognized and elaborated. In his development of differential geometry (a subject that had been created several years earlier by Riemann's mentor, Carl Friedrich Gauss), Riemann used the language of Descartes's coordinate definition of dimension, but, rather than ignoring the implication of a mathematical fourth dimension, he developed it in detail.

Riemann explicitly proved that there were other geometries besides Euclid's that described worlds of every whole-numbered dimension, from zero to infinity. The 3-D world described by Euclid was now openly recognized to be only one of many equally logical possibilities and to be a rather prosaic one at that.

More things are possible in a 4-D world than in a 3-D world. To a person in a 4-D world, the goings-on of a 3-D world would look as flat and confined as the action on a 2-D movie screen looks to us. Conversely, the fourth dimension is as imperceptible to a 3-D person as the third dimension is on a 2-D movie image. Anything that would enter a 3-D world from the fourth dimension would seem to come from nowhere (shades of Henry More's ghosts!). This means that 4-D persons could observe a 3-D world unnoticed, just as if they were spectators at some movie house. To create the illusion of depth, they would merely need to don a pair of 4-D glasses.

Other worlds described by Riemann's mathematics were not even spatial in the ordinary sense, as were Euclid's and Descartes's. According to Riemann, mathematical dimension need not refer only to sensible space; it could just as logically refer to purely conceptual spaces, which he named manifolds. In taking this imaginative leap toward abstraction, Riemann liberated geometry even more than Descartes had from Euclid's dependence on the physical notions of length, width, and depth.

Although it has acquired some additional technical connotations since then, a manifold still refers generally to any realm—whether it is the stock market, the economy, or a human being—whose description is multifaceted, or multidimensioned, as Riemann preferred to call it. Whereas to previous generations of geometricians "space" had referred exclusively to that familiar realm filled with galaxies, stars, and planets, to today's geometricians, the word "space" has expanded to mean "manifold."

It should be stressed that Riemann's was not simply a semantic revolution. In referring to the stock market, for instance, as a many-dimensioned manifold, mathematicians think of it literally as a geometrical space in which things behave according to certain mathematical theorems. The dimension of the geometrical space corresponds precisely to the number of factors that govern the stock market's behavior. If we imagine for the moment that the stock market depends on but one factor—the hemlines of women's skirts, say—then we would represent it mathematically by a 1-D manifold, a line. This means, among other things, that the status of the stock market at any time would be described by a single number (the height of the hemlines) and illustrated as a point somewhere along the 1-D manifold.

Thought of in this way, a human is a manifold of an extraordinary number of dimensions—some might even say an infinite number. Typically, our behavior is influenced by an inestimably large variety of factors. Even behavior as elemental as deciding whether to smile at a stranger may depend on such far-flung factors, or dimensions, as the amount of sleep we had the night before, the time of day, our feelings toward our spouse at the moment, our job security, the state of the economy, the season of the year, the stranger's appearance, and on and on. And it is not inconceivable that some of our most complex behavior can depend on the detailed state of the entire universe.

So pity the social scientists. One of the reasons that their track record looks so miserable compared to that of the physical scientists is simply because their job is more difficult, and perhaps even impossible. If we do think of a human being as an extraordinarily high-dimensioned geometrical world, then trying to fully describe it mathematically and scientifically is a challenge that dwarfs similar efforts to describe our 3-D spatial universe.

Indeed, the physical scientists' voluminous receptacle, the universe, is one of the simplest, least-dimensioned manifolds around. According to Einstein, it is a 4-D manifold, not a 3-D one, as everyone once perceived it to be. In 1915, taking advantage of Riemann's liberal definition of dimension, Einstein concluded that the universe appears to be most accurately modeled after a geometrical world that has three spatial dimensions and one temporal dimension.

It is worthwhile to emphasize that such an assertion would have been without mathematical meaning had Riemann not generalized the definition of dimension beyond the purely spatial connotations of length, width, and depth. A temporal dimension has none of these qualities; that is, time is not measured with a spatial ruler.

According to Einstein, the universe is a 4-D space-time manifold, because (in Riemann's language) three space coordinates and one time coordinate are required to map it. Each of its points is uniquely labeled by three numbers that specify position and one number that specifies time. This is a technical way of saying that if we wish to meet someone, we must specify both a place and a time. The place always contains three separate pieces of information (for example, 5th Avenue and 56th Street, on the surface of the earth) and the time, one piece of information (say, noon). In a sense, then, Einstein proved that Ptolemy was not entirely incorrect: the *spatial* part of the universe is such that it can accommodate only three coordinate axes.

Because one of its four dimensions is time, the universe according to Einstein is not like the purely spatial 4-D world described earlier. In that world, there is no time and therefore no concept of movement as we know it. A person in a purely spatial 4-D world exists at all locations simultaneously. In our 4-D space-time universe, however, a person is always in one place at any given time and is restricted in his movements from one place to another by his maximum speed of travel. (The absolute maximum speed limit in the universe, scientists have found, is the speed at which light travels in a vacuum—about 186,000 miles per second. We, of course, are restricted to much smaller speeds.)

Though it took us well over two centuries to recognize that the universe is four-dimensional and contains things that have dimensions of various kinds and numbers, it has taken us another sixty years or so to recognize that the dimensions of many other things within the universe are not even whole numbers. And once again, it was a mathematician who called this possibility to our attention.

In 1975, Benoit Mandelbrodt of IBM consolidated and reinterpreted the desultory work of many earlier mathematicians to show that it is mathematically feasible to define a fractional dimension such as ¾-D, 1½-D, and so on.

In developing his thesis, Mandelbrodt began with a definition of dimension proffered by the German mathematician Felix Hausdorff some six decades after Riemann had articulated his. According to Hausdorff, an ordinary surface, such as this page, is two-dimensional because we need to multiply two numbers (that is, length and width) to calculate its area. Similarly, an ordinary solid, such as a sugar cube, is three-dimensional, because we need to multiply three numbers, (length, width, and depth) to calculate its volume, and so forth. Hausdorff believed that by following this simple rule, it would be possible to catalog all conceivable geometric figures, from zero-D to infinity-D.

Mandelbrodt observed, however, that mathematicians in the past had conceived of figures that would defy being cataloged in this way. He offered several of his own examples of such figures, based on common natural phenomena such as the coastline of Britain: Imagine a rectangle whose sides all around are as ragged as the coastline of the island of Great Britain. According to Hausdorff, such a rectangle is 2-D because its area is calculated by multiplying its length by its width. However, Mandelbrodt notes, Hausdorff's prescription in this case is more easily said than done; there is a difficulty in deciding what the proper length and width of the rectangle is, and this leads to a paradox that is the hallmark of fractionally dimensioned objects.

Seen from a distance, the sides of the rectangle appear smooth, and so we might be apt to say that the length and width are, respectively, ten and twenty miles long "as the crow flies." According to Hausdorff's prescription, the area of the rectangle is 200 square miles.

Seen closer up, however, the rectangle's jagged edges become readily apparent. If we were to follow their every cove and peninsula by automobile, we would find that the sides are considerably greater than ten and twenty miles long. Judging from a reading of the automobile's odometer, we might find that the width is closer to fifteen miles and the length is more like thirty miles. Clearly, this is a more accurate estimate than the first one, and yet, if we followed Hausdorff's prescription, we would be led in the second instance to grossly overestimate the rectangle's area at about 450 square miles.

The estimate of the rectangle's length and width made from an automobile is even less accurate than the estimate made by a small bug following every minuscule projection and cove of the ragged coastline. "In principle," Mandelbrodt says, "man could follow such a curve down to finer details by harnessing a mouse, then an ant, and so forth. As our walker stays increasingly closer to the coastline, the distance to be covered continues to increase with no limit." The paradox inherent in Hausdorff's dictum, then, is that as our estimate of the length and width improve, our estimate of the area worsens.

Examples like this one, which do not fit nicely into Hausdorff's cataloging scheme, indicated to Mandelbrodt early on in his studies that Hausdorff's definition of dimension had to be expanded somehow to include them. It was his effort to accomplish this purpose that led him to fill in the gaps of Hausdorff's dimensional hierarchy with a mathematical theory of fractional dimensions.

Mandelbrodt found that in the case of a geometric object such as a rectangle, the paradox vanishes if we do not insist that the area is the product of two numbers. According to his theory, the correct, paradox-free prescription for finding the rectangle's area is to multiply together not two numbers but one number times the square root of another number. This means that the rectangle is effectively a 1.5-D object. (Here, one half of a number refers to the square root of a number; similarly, one-third would refer to the cube root, and so forth. Therefore, multiplying together 1.5 numbers means multiplying one number times the square root of another.) These numbers still have something to do with length and width, but they have been mathematically redefined by Mandelbrodt so as to be unaffected by the point of view we choose to adopt, whether it is cosmic, particular, molecular, atomic, subatomic, or anything smaller.

The example of a rectangle with sides like the British coastline illustrates how a planar object can confound Hausdorff's whole-number dimension scheme; a cube with its surface as convoluted as the human brain illustrates the same thing for solid objects. According to Hausdorff, such a cube is 3-D because its volume is calculated by multiplying together three numbers—its length, width, and depth. But, as with the rectangle, we find that the most detailed measurement of these three quantities—that is, one that follows every tiny convolution—leads us to greatly overestimate the volume of the cube. Mandelbrodt found that such a cube is a fractionally dimensioned object, a fractal, whose volume is the product of 2.8 numbers, not three, and whose dimension, therefore, is 2.8.

Mandelbrodt found many other familiar objects to be fractals. Planar fractals include snowflakes and the outlines of mountain ranges, and solid ones include automobile radiators and human intestines. For each of these, as for the rectangle and cube, Hausdorff's prescription leads to a paradoxical relationship between the object's area or volume and the lengths of its sides.

Mandelbrodt also found that the typical fractal dimension of Earth's natural landscape is different from that of Mars's, judging from NASA photos of the red planet: Earth is somewhere around 2.1 and Mars is about 2.4. To the eye the main difference is that the jaggedness of the terrestrial landscape is somewhat less "toothy" than the Martian one. One practical consequence of this observation is that any possible landscape, each one corresponding to a particular fractal dimension, can now be generated by computer. In fact, this is exactly what the producers of the 1981 Paramount film *Star Trek II: The Wrath of Khan* did to create alien-looking skylines for various fictional planets. With his theory of fractals, therefore, Mandelbrodt has allowed us to appreciate not only the dimensional richness of our local world, but also its uniqueness.

With their novel insights into the mathematical nature of dimension, Mandelbrodt, Riemann, and Descartes have enabled us to perceive the universe as texturally richer than the rather plain 3-D spatial receptacle filled with only the 1-, 2-, and 3-D objects of Euclid's day. It is tempting to think that the future may be an extrapolation of the past, and that we will continue to perceive more dimensions in the texture of the universe.

It might also be imagined, however, that in the future the texture will appear to wax and wane, as though the universe were a ball of string through which we are traveling. The possibility conjures up images of the movie *Fantastic Voyage*, in which several scientists were shrunk to a very small size to explore the interior of a human body.

Our earliest view of the universe can be likened to looking at a ball of string from a long way away—it looks like a zero-dimensional point. From closer up it appears to be a two-dimensional disk with a smooth surface; closer still, and the 2-D surface appears to have some 3-D texture. This last may be the point at which we are now in our perception of the universe, a point at which the full richness of the surface texture is most apparent to us. If so, then we are heading toward a moment that corresponds to breaking through the surface of the ball of string, when the appearance of richly dimensioned texture gives way to the appearance of a space sparsely populated by one-dimensional threads. Not until we head toward one of these threads will we once again begin to perceive the 2-D, then the 3-D, texture of its fuzzy strands.

We cannot predict with any certainty, of course, which of these future possibilities will come to pass. The question remains whether the number of dimensions we impute to the universe and its contents will continue to increase steadily or fluctuate. It is certain, however, that the universe, like that businessman accused of being one-dimensional, must not be judged hastily and with any finality. For if that distinguished line of inspired mathematicians has taught us anything in the last 2000 years, it is that dimension is in the eye, and in the imagination, of the beholder.

Related Essays

Nothing Like Common Sense
The Familiar Faces of Change

Zero and the Null Set

MUCH ADO ABOUT NOTHING

> Thirty spokes share one hub. Adapt the nothing therein to the purpose in hand, and you will have the use of the cart. Knead clay in order to make a vessel. Adapt the nothing therein to the purpose in hand, and you will have the use of the vessel. Cut out the doors and windows in order to make a room. Adapt the nothing therein to the purpose in hand, and you will have the use of the room. Thus what we gain is Something, yet it is by virtue of Nothing that this can be put to use.
>
> Lao Tzu

Mathematicians have the distinction of being among the first people in history to recognize the value of nothingness and to understand and express the difference between nothingness and nothing. Even though these may sound like vacuous milestones, they were prodigious events in the development of arithmetic (the study of numbers) as we know it today. Between eleven and fourteen hundred years ago, nothingness found its permanent place in arithmetic as the number zero. It was a seemingly trivial addition to our numerical alphabet, but it enabled us to express numbers more clearly and easily than ever before. Thirteen hundred years later, nothing made its appearance in arithmetic, and it was not anything like nothingness. Nothing was named the null set, and it represents in arithmetic what a virgin canvas and a blank sheet of paper represent in painting and writing: latent being. With the null set, mathematicians have been able to show how every number known in arithmetic can be created out of nothing. To paraphrase Shakespeare, never before or since have mathematicians made so much of so little as when they made much ado about nothing and nothingness.

The Hindus permanently introduced the zero, which they called *sunya* (the void), into their number system sometime between the sixth and ninth centuries A.D. Today's number system is a direct descendant of that early one, and is known to mathematicians as "positional notation decimal" (PND). In this system, any number, however large, can be expressed using only the ten basic digits, zero through nine. Furthermore, the position of each digit in a number, such as 8754, determines its actual value. By convention, the rightmost digit (4) stands for itself, the next digit to the left (5) stands for ten times itself, the next digit to the left (7) stands for one hundred times itself, and so forth.

Before the sunya came into being, mathematicians who used positional notation systems normally left blank spaces where we would now write in zeros. For instance, the number seven hundred seven would have been written 7 7, rather than 707. Often, the spaces were not made clear enough, and this led to an ambiguity that was not only a nuisance but that also impeded the development of arithmetic.

There were several other kinds of number systems being used at the time the Hindus came out with theirs, but none of them used zero and none of them made it as easy to express numbers. There was, for instance, the Roman numeral system. It was unwieldy and complicated, as anyone who has ever tried to use it knows. It is easy enough to use for small numbers, but for large numbers, the Roman numerals take on the aspect of a formidable puzzle. For example, the Roman cipher for one thousand, nine hundred eighty three is MCMLXXXIII—compare this with the much simpler Hindu 1983!

The Greeks, too, had a number system similar in operation to the Roman numeral system. In order to express the number 1983, they had to write TΣTNIIIAAA. It is not surprising, then, that arithmetic flourished with the sunya-conscious Hindus, but not with the Greeks, who are remembered today for their prowess in geometry (the study of shapes). This is no more surprising than that the development of poetry (to which mathematics as an art form has often been compared) in a culture has usually depended on the expressiveness of that culture's language.

If the mathematical concept of nothingness has enabled us to express numbers more concisely and accurately, then the mathematical concept of nothing has provided us with a means to help elucidate the logical origins of numbers.

Around 300 B.C., when Euclid used logic to derive geometrical truths from a small number of assumptions, the question was raised of whether the same could be done with arithmetical truths. When mathematicians finally applied themselves to find the answer to this in the late nineteenth century, they could not agree on where to start. Some wanted to begin by merely presuming the mathematical existence of the sequence of natural numbers (that is, 0, 1, 2, 3, . . .), the basic alphabet of arithmetic, and go on from there. Others, like the German Gottlob Frege, wanted to begin further back logically, to derive the sequence of natural numbers itself from some even more primitive concept or concepts.

As one of his logical starting points, Frege chose the common-sense notion of a class, or set. In mathematics, as in any other application, a set is simply a group of things; or, as one of Frege's contemporaries, the German Georg Cantor, put it, a set is "any collection . . . of definite and separate objects of our intuition or thought." Frege believed that the idea of a set was an even more primitive notion than the sequence of natural numbers, and so he proceeded from it to derive the sequence; from there, using additional assumptions, he could derive the whole of arithmetic. In the process, he recognized the difference between nothingness and nothing.

The difference, he discovered, is one of class. Nothing is the empty, or null, set—the set with no members—which mathematicians denote with a pair of braces: { }. It was also the only possible set that Frege had to begin with, since he had started out by not assuming the existence of any numbers. For him, the null set represented the moment just before creation, the potential for becoming an infinite sequence of numbers. Even the symbol for it suggested latent being: two braces enveloping an emptiness that would eventually be filled by a growing sequence of numbers.

Nothingness, by contrast, is the set with zero as its only member, which mathematicians write as {0}. Just by looking at this symbol, we can see what Frege saw: nothingness is not nothing, but is actually something. In Frege's analysis, {0} was the first conceivable realization of the null set's mathematical potential, the first natural number, zero. The set {0} is the blank realization of the null set, just as a musical pause is the blank realization of a musician's potential to create sound.

It is less easy to recognize the difference between nothingness and nothing in areas outside of mathematics. For example, when I stared recently at the celebrated Henry Moore sculpture at the entrance of the east wing of the National Gallery of Art in Washington, D.C., I saw in its single, central opening the potential for many images. It was at once nothing and everything, and in that sense it was a model of the mathematician's null set. But just as often, in the voids of other sculptures, I have seen models not of nothing, but of nothingness—just holes.

One of the truest physical analogues of the null set is the physicist's theoretical notion of a vacuum. It is a quintessential example of corporeal nothing, imputed with a potential for giving birth to prodigious quantities of matter. Theoretically, all that is required to induce this spontaneous creation of nuclear particles is that the vacuum be bathed in a very strong electrical field.

So far, this modern portrayal of a vacuum has been impossible to test directly, because the strength of the required electrical field exceeds what we are able to generate artificially. Recently, however, a group of nuclear physicists and chemists have proposed testing the theory by marshalling the naturally superstrong electrical fields of what they call superheavy atoms. The only problem with this idea is that at present the existence of these superheavy elements is also a matter of conjecture. Thus, we still cannot be certain whether the vacuum is actually nothing, as today's physicists believe, or nothingness, as many of the ancient Greek philosophers believed.

As striking an analogue of the null set as the physicist's vacuum is the biblical account of divine creation. According to the Bible, God created the universe just as Frege created the sequence of natural numbers—out of nothing. In her play *Sapientia*, the ninth-century nun playwright Hrovita of Gandersheim goes even further: "[T]he Author of the world . . . created the world out of nothing, and set everything in number, measure and weight, and then, in time and the age of man, formulated a science which reveals fresh wonders the more we study it."

I'd like to believe that Sister Hrovita would include mathematics in her "science" and Frege's use of the null set in her "fresh wonders." If she would, I believe she might also include as a fresh wonder a recent reapplication of the null set to arithmetic by the English mathematician John Horton Conway. In 1973, he picked up where Frege left off by beginning with nothing, the null set, and creating from it not only the natural numbers and also the other known kinds of numbers, the fractions and the nonfractional decimals (otherwise known as the irrational numbers), but also some heretofore unknown kinds of numbers, called surreal numbers.

In his thirteen-page manuscript, "All Numbers, Great and Small," Conway begins as Frege began, with a few primitive ideas, including the null set and two rules. The first rule, Conway's logical definition of a number, can be visualized in terms of encyclopedia volumes lined up in order on a library shelf. According to the definition, a volume's place in the lineup, its number, can be inferred from the set of volumes on its left and the set of volumes on its right. We could determine where volume nine belongs, for instance, simply by locating that place where volumes zero through eight are on the left and volumes ten through infinity are on the right. Therefore, every volume, every number, has its own niche, determined uniquely by the left and right sets. That's the thrust of Conway's first rule.

His second rule, again explained here in terms of a set of encyclopedias, decrees that one number, such as 5, is smaller than (or equal to) another number, such as 9, if two things are true simultaneously: (A) all the volumes to the left of the first number (5) are less than the second number (9), and (B) all the volumes to the right of the second number (9) are bigger than the first number (5). This rule is necessary in order for Conway to impose an order on the numbers he creates, beginning with zero: Zero is less than 1, so it precedes 1; 1 is less than 2, so it precedes 2; and so forth.

As he does not assume the existence of any numbers to begin with, Conway, like Frege, has only the null set with which to start creating the sequence of natural numbers. Consequently, Conway first contemplates the number whose left and right sets are both null sets, written symbolically as { }:{ }. He names this *zero*. That is, in Conway's theory, as in Frege's, nothingness is the most primitive realization of nothing.

After creating the number zero, Conway has two sets

with which to continue creating numbers: the null set, { }, and the set containing zero, {0}. Conway identifies the number 1 as the number whose left set contains zero and whose right set is the null set. Thus, at this point in Conway's genesis, the number 1 is flanked to the left by nothingness and to the right by nothing. To the left is potential already realized (as zero), and to the right is potential not yet realized.

At each point in his creation, Conway always selects the next number as the number whose left set contains all the previously created numbers and whose right set is the null set. It's as though he were being guided by an image of those encyclopedias. At each point, the newly created volume is placed to the right of all those volumes already shelved and to the left of empty space, which in this analogy has the aspect of the physicist's vacuum in representing the potential of numbers not yet brought into being. By proceeding in this fashion indefinitely, Conway creates the entire sequence of natural numbers.

From there he goes on, however, to create an infinity of in-between numbers, such as the number whose left set contains zero, {0}, and whose right set contains one through infinity {1, 2, 3, . . .}. This defines a number somewhere between zero and one. Thus the standard set of encyclopedias, the natural numbers, is embellished by an interminable number of in-between volumes. And it doesn't stop there.

Pursuing the logic of his method, Conway is able to create between in-between numbers, then numbers between *these*, and so on, literally ad infinitum. The result is limitless hierarchies of in-between numbers, never before named in mathematics.

Conway's theory has ineffable graphic implications as well. Traditional mathematical wisdom has it that a ruler's edge, a number line, is a blur of points, each of which can be labeled with either a whole number, a fraction, or an irrational number such as .1345792 . . . , where the string of digits goes on forever. All these points (or their numerical labels) together are imagined to form a continuum, with no space between adjacent points. Conway's theory, however, asks us to imagine numbers that fall somehow between unimaginable cracks in this blur of points, and between the cracks left behind by those numbers, and so on and so on. With his theory, Conway has made credible what many persons before him had merely speculated about: there is conceptually no limit to how many times an object can be divided.

Conway's "All Numbers, Great and Small" shows off the boundless potential of the null set, but also of the human mind. Human creative energy, like nothing, isn't anything if it isn't potential. It is also an indomitable part of being alive, as countless experiments have documented. People who are deprived of their senses by being floated in silent, dark tanks of water warmed to body temperature will hallucinate. It is as though the human mind will not be stilled of its propensity to make something of nothing even, or especially, when immersed in nothingness.

Like a physicist's vacuum, the human mind can be induced to create thoughts that come seemingly out of nowhere. Mathematicians over the years have documented this common phenomenon. The German Carl Friedrich Gauss recalled that he had tried unsuccessfully for years to prove a particular theorem in arithmetic, and then, after days of not thinking about the problem, the solution came to him "like a sudden flash of lightning." The French mathematician Henri Poincaré, too, reported working futilely on a problem for months. Then one day while conversing with a friend about a totally unrelated subject, Poincaré recalled that ". . . the idea came to me without anything in my former thoughts seeming to have paved the way for it."

In this sense, the human mind is the real null set in Frege's and Conway's number theories; the mathematical null set is but a subordinate entity created after the mind's self-image.

And if Frege's and Conway's creative experiences were anything like those of Gauss and Poincaré, then it was by thinking of nothing in particular that they advanced the subject of arithmetic. It is to their minds, in particular, that we owe this means of recognizing the enormous difference between nothingness and nothing.

Related Essay

A Certain Treasure

PART TWO

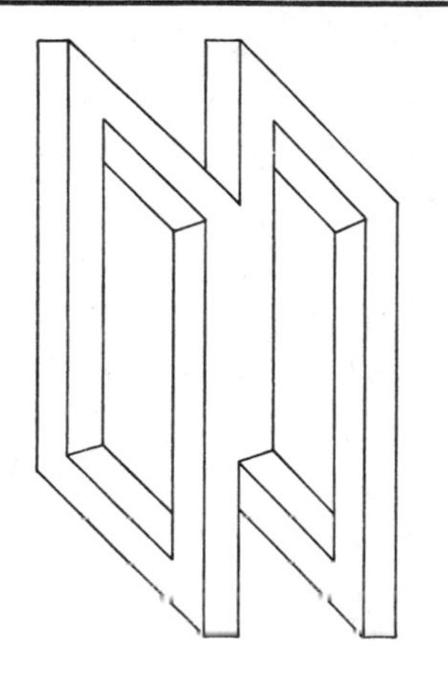

COMPROMISING

Non-Euclidean Geometries

NOTHING LIKE COMMON SENSE

> A group of hunters, having pitched camp, set out to go bear hunting. They walk one mile due south, then one mile due east, at which point they sight a bear. Bagging their game, they return to camp and figure out that altogether they have traveled three miles. What was the color of the bear?
>
> American riddle

Our tendency to judge the natural world in terms of our common sense has led us at various times to believe that the sun actually rises and sets, that the stars revolve around the earth, that the earth is at the center of the universe, and that the earth is flat. All these anthropocentric beliefs died hard, but none harder than the general belief that the universe was in some ways merely an extension of the terrestrial environment.

Until the early nineteenth century, it was believed that the laws of Euclid's geometry were obeyed in every region of the universe exactly as they were here on earth. It was a belief that made for a cozy feeling about the cosmos, and when it was finally discredited, the universe suddenly lost its familiarity—and common sense again lost its credibility as a reliable judge of things.

The source of this discreditation was a discovery in 1824 that there were geometries other than Euclid's that were just as mathematically valid, but that described universes that were dramatically dissimilar in some ways to the universe described by Euclid's geometry. With this discovery, the possibility was created that our universe was not Euclidean after all, but that it was like one of those portrayed by the new geometries.

As a result, the universe became an unknown quantity, and in the following decades our understanding of it had to be formulated anew. This was largely accomplished in the 1920s through the ideas of Einstein's general theory of relativity. This time our relationship with the universe was based not on common sense but on the *uncommon* sense provided to us by modern geometry.

The geometry that would eventually be identified with Euclid originally came together in ancient Egypt as bits and pieces of practical knowledge about land surveying and architecture (the word geometry comes from the Greek for "to measure the earth"). The seminal ideas about points, lines, planes and solids were shaped from common-sense ideas formed from experiences with landmarks, footpaths, farmers' fields, and granite blocks. The important geometrical concept of parallel lines as lines that never meet was no doubt firmly associated in Egyptian minds with such earthy phenomena as plow lines or the ruts dug into the roads by two-wheeled carts. In view of its origins, geometry should have been thought of merely as a collection of mathematicized truisms about how points, lines, planes, and solids behave here on earth, but in fact it was thought of by the Egyptians, and then by the Babylonians and the Greeks, as a collection of truisms about how points, lines, planes, and solids behave not only here but throughout the universe.

By Euclid's time, around 300 B.C., astronomers were

using the theorems of geometry as if they were scientific laws. If they were to imagine, as the astronomer Eudoxus did, that the stars moved on a huge crystalline sphere, they automatically assumed that a sphere in space is just like a sphere on earth. And if they imagined the distance between two heavenly objects, they took it for granted that in space the shortest distance between two points is a straight line, just as it is on earth (which they assumed was flat). In short, in imagining the geometry of the universe, Greek astronomers nonchalantly applied their terrestrial experiences to an arena much larger than the earth.

In a way, Euclid's contribution to geometry only served to intensify the belief in geometry as universal science. What he did was to show how the hundreds of geometrical theorems that had accumulated over the centuries could be logically derived from just ten postulates. Among the ten were such apparently universal truisms as "A straight line can be drawn from any point to any other point," "All right angles are equal to each other," "If equals are added to equals, the sums are equal," and "The whole is greater than the part." Euclid's was a prodigious, unprecedented achievement in mathematics, and it lent to geometry an aura of universal and irrefutable verity.

The only challenges to Euclid's compelling enshrinement of common sense were directed at his second and fifth postulates. These postulates were, respectively, "A finite straight line can be extended indefinitely to make an infinitely long straight line" and "Given a straight line and any point off to the side of it, there is, through that point, one and only one line that is parallel to the given line." In Euclid's day and for centuries thereafter, most mathematicians expressed varying degrees of doubt that these were genuine postulates. It wasn't that the mathematicians doubted that they were true—that seemed to them to be indicated by common sense—but they did not agree with Euclid that the two postulates were self-evident truths. Instead, the skeptics believed, Euclid's second and fifth postulates were actually theorems that could logically be derived from the other eight postulates.

The reticence to accept these two postulates as self-evident truths was based mostly on various hunches of some mathematicians. Some could not bring themselves to avow the self-evident truth about anything having to do with infinity; others were suspicious about the fifth postulate in particular because it expressed an idea that struck them as somehow more involved than the others. For whatever reasons, all these skeptical mathematicians were reluctant to accept Euclid's second and fifth postulates *on faith*; the skeptics required proofs of the postulates' veracity, but at the same time they never seriously doubted that someday such proofs would be forthcoming.

Those expected proofs never came, however. Instead, one day in 1824, the eminent German mathematician Carl Friedrich Gauss received a letter from an old school friend of his, Farkas Bolyai. Bolyai, a mathematics teacher, was writing to ask Gauss to evaluate an accompanying manuscript written by Bolyai's son, János. The young Bolyai had made an astonishing discovery that appeared to resolve at long last the uncertain status of Euclid's fifth postulate.

First of all, János had proved that the parallel lines postulate *was* actually a postulate. If it was to be a part of Euclid's geometry at all, then it had to be accepted on faith, as a self-evident truth.

János also demolished the long-cherished belief in the universal verity of Euclid's geometry. He did this by first replacing Euclid's parallel lines postulate with another parallel lines postulate that appears contrary to common sense: "Given a straight line and any point off to the side of it, there is, through that point, an infinite number of lines that are parallel to the given line." Then, from this and from Euclid's nine remaining postulates, János derived theorems that were dramatically different from those of Euclidean geometry but that were just as logically well-founded. They described, as he later put it, "a new universe." Because of the parallel lines postulate in particular, the geometry of this "new" universe was not anything like that of our terrestrial environment.

Gauss read the manuscript with great interest and a sense of familiarity, for in fact he himself had made these same discoveries a few years before. He revealed this in a letter to the elder Bolyai, explaining that he had kept quiet about the discovery for fear of the outspoken displeasure it was likely to provoke among his colleagues. In those days, Euclid's geometry was still as revered as the Bible; the discovery just described was akin to a discovery of a second Bible whose description of Christianity was dramatically different from that of the first.

In 1832, the Russian mathematician Nikolaus I. Lobachevski independently made the same discovery that János Bolyai and Gauss had made. Apparently, this news, combined with Gauss's earlier disclosure, so dispirited the young Bolyai that he eventually forsook a career in mathematics and joined the cavalry.

When mathematicians first learned of the discovery, they reacted with emotion, just as Gauss had foreseen. The initial reaction was disbelief: it did not seem possible that (1) Euclid's geometry was merely one kind of geometry and not the geometry of the universe, nor that (2) a mathematically sensible geometry could be derived from a postulate that was apparently not a self-evident truth.

The second point was especially difficult for mathematicians to grow accustomed to. Its chief implication is that mathematics is a mere invention of the human imagination and not a body of universal truths based on common sense, as everyone had believed. If someone such as Bolyai, Gauss, or Lobachevski could simply invent a postulate that did not necessarily have anything to do with reality—and certainly it had nothing to do with common sense—and then proceed to use it to derive a logically valid mathematical system, then mathematics itself was an invention. In particular, Euclid's geometry was an invention, albeit one that was based more on common-sense principles than was Lobachevski's (as mathematicians now simply refer to the three-way discovery).

By the late 1800s, mathematicians were all fairly resigned to the discoveries that had so thoroughly disabused them of the traditional view of Euclid's geometry and of mathematics generally. They even began to take in stride the possibility that there could be more than one mathematically valid geometry, and that it was a question best left to science as to which among the variety of invented geometries, if not Euclid's, actually does describe the universe.

By that time, mathematicians were already aware of the three basic varieties of geometry known today. Besides Euclid's and Lobachevski's geometries, there is also Bernhard Riemann's geometry, named after the German mathematician who invented it in 1854.

Riemann's geometry differs from Euclid's in the second and fifth postulates, in both cases in ways that defy common sense. Riemann's second postulate, "A finite straight line cannot be extended indefinitely to make an infinitely long line," is literally the logical opposite of Euclid's. Riemann's fifth, or parallel lines, postulate is unlike both Euclid's and Lobachevski's. It reads: "Given a straight line and any point off to the side of it, there are, through that point, not any lines that are parallel to the given line."

Because Lobachevski's and Riemann's geometries are each based on at least one postulate that is inconsistent with our common-sense, earthbound experience, they describe the behavior of points, lines, planes, and solids in worlds that are unfamiliar to our senses. Nonetheless, it turns out that these two bizarre, non-Euclidean worlds can be pictured in terms of two familiar figures.

A world in which Lobachevski's geometry describes reality is like the surface of two infinitely long heralder's trumpets with their bells butted together. This surface is called a pseudosphere, but it looks nothing like a sphere. On such a world, straight lines are lines that run lengthwise along the trumpetlike surface. With some concentration, and in terms of this model, it is possible to see just how Lobachevski's peculiar parallel lines postulate can be obeyed in a sensible way. That is, it is possible, given the peculiar shape of the trumpetlike surface, to draw an infinite number of distinguishable lines through a single point so that each of those lines is parallel to some given line.

It is easier to picture the world of Riemann's geometry. Such a world is like a sphere, and a straight line is like the arc of a great circle (the arc and straight lines are true counterparts, because on a sphere the arc is the shortest distance between two points, just as a straight line is on a flat surface). It is easy to see, in terms of this model, just how Riemann's two counterintuitive postulates can be sensibly obeyed. In the case of his second postulate, straight lines cannot be extended indefinitely to make an infinitely long line, because they are like arcs of a great circle that have a finite length determined by the sphere's diameter. In the case of his fifth postulate, parallel lines cannot exist in a Riemannian world because arcs of great circles always converge.

The world in which Euclid's geometry describes reality can also be likened to a surface, the ordinary plane. In such a world—the world most familiar to our senses—straight lines are just straight lines upon the plane. It is especially easy to see, in terms of this model, just how Euclid's parallel lines postulate is obeyed: On a flat plane surface, no more than one line can be drawn through a point that will be parallel to some given line. Also, on a plane surface, straight lines can be extended indefinitely to make an infinitely long line.

The question that mathematicians felt was best left to science—that of whether the geometry of the universe is like that of a pseudosphere, a sphere, or a plane—was not given much attention by scientists until well into the twentieth century. The main reason for this inattentiveness to a question of such major significance was that it took some time for scientists to realize that it is scientifically feasible to answer it.

This was not an automatic realization, as it was not obvious at first just how we could go about determining the geometry of something of which we can only see a small part. It seemed about as feasible for us to evaluate the geometry of the universe as it would be for a barnacle to evaluate the geometry of the whale to which it is attached. The scientist who first realized that the question was, indeed, answerable and who showed us how it could in principle be answered was Albert Einstein.

In 1915, Einstein published his general theory of relativity, which explained how it is possible to infer the global geometry of something as big as the universe by piecing together observations done over distances as relatively microscopic as those presently accessible to human beings. The gist of its explanation is illustrated by the following situation. Imagine yourself to be a cartographer who is asked to infer the shape of the earth by hiking around some part of its surface. Equipped with the very best in instruments, you begin by walking 100 miles successively in three directions: south, west, and north. You can foresee that if you now walk for exactly 100 miles due east, you will end up right where you started, having traversed all four sides of a 100-mile square. So, with that expectation in mind, you set out on the last leg of your hike but reach home more quickly than you had expected. Your measuring instruments record less than 100 miles—they indicate something closer to 99 miles. Either your starting point has moved while you were away, or the accuracy of your instruments is in question. What gives?

Your common sense is what gives: It misled you to assume a flat earth on which the four cardinal directions are exactly perpendicular to one another. Your starting point has never moved, and your instruments are fine—indeed, they are letting you know that the earth is not flat. You have not traversed a perfect square, as you had assumed, because the earth is roughly spherical; consequently, its geometry is not Euclidean but Riemannian.

Here's what actually happened to you on the hike. Beginning somewhere in the northern hemisphere, you headed south toward the equator along a longitudinal line. After 100 miles, you hiked westward along a latitudinal line for 100 miles, then north toward the pole along another longitudinal line for 100 miles. Your return home from that point was less than 100 miles simply because longitudinal lines converge as they get closer to the pole (and diverge as they get closer to the equator).

(You might now have figured out that the color of the bear in the opening riddle must be white, because the hunters' triangular path makes sense only if one of its vertices, the campsite, is on the north pole.)

Today, astronomers hope to infer the geometry of the universe in the same way that we have just hypothetically inferred the geometry of the earth—that is, by mapping some part of it with great accuracy. They are sanguine about their chances, partly because of their confidence in Einstein and his theory, and partly because their own understanding of the various geometries has become so thorough that "uncommon" sense has now become incorporated into our thinking. Through uncommon sense we can extend our feelings of intimacy with things, including hypothetical worlds, that are physically inaccessible to us and thus alien to our *common* sense.

Armed with optical and radio telescopes and with Einstein's theory, modern cosmologists are surveying the local distribution of matter—landmarks, as it were, chiefly comprised of stars and diffuse clouds of luminous gas and dust. Their intention is to decide the question of the geometry of the universe. So far, their observations have not been precise enough to produce an unambiguous answer, but it is beginning to look more and more as though the geometry of the universe is like that discovered by Gauss, Bolyai, and Lobachevski.

In a sense, this tentative scientific result is the second punch in a one-two combination directed against our historical exaltation of Euclidean geometry. In the wake of these and numerous other blows dealt against common sense over the last several millenia, we might well wonder whether we can trust our common sense at all in studying the far-flung universe. In short, is our common sense in any way universal?

Our awareness and understanding of non-Euclidean geometries has enabled us to answer that question. What mathematicians now realize is that although the three worlds associated with the three possible geometries would look very different from one another when seen from a bird's-eye view, they become indistinguishable when seen from an ant's-eye view.

This observation is easy to picture in terms of those model surfaces. When looked at from a wide perspective, over a large distance, the pseudosphere, sphere, and plane cannot be confused with one another. But when looked at from a very narrow perspective, over a relatively small distance, all three look alike—they all look like planes.

This fact should not surprise us, since it is consistent with the reasons our predecessors misjudged the earth's curved surface as flat. But the mathematical relationship might well have been some other way around—it could have turned out that all geometries when seen up close look Riemannian or Lobachevskian. In fact, though, it has turned out that Euclidean geometry alone has this role in mathematics, and this clarifies something about the origins of our Euclidean common sense.

It is all a question of scale; our Euclidean common sense has nothing to do with our humanity nor with our human environment per se. The common sense that induced us to believe in the universality of Euclid's geometry 2000 years ago depends only on the scale of our existence and the perspective of our senses, not on any unique anthropocentric tendencies. Perhaps there are other creatures living elsewhere in the universe who, sharing our restricted perspective, can be expected to have also revered Euclidean geometry sometime during the beginning of their intellectual histories.

Ironically, then, because of the uncommon sense that we have acquired in learning about non-Euclidean geometries, our common sense has at once been humbled and exalted. It has been humbled, because we now know that it is an unreliable guide to our understanding of the world and beyond. But at the same time it has been exalted, because we now know with mathematical precision that if extraterrestrial life is as prevalent as many astronomers believe it to be, our common sense is positively universal.

Gödel's Theorem

AN ARTICLE OF FAITH

What is laid down, ordered, factual, is never enough to embrace the whole truth.

Boris Pasternak

Until about fifty years ago, *truth* to a mathematician had been synonymous with logical proof. An hypothesis was true if it could be proved with logic and false if it could not be. For this reason, mathematicians had operated in a fantasy world, one in which nothing was left to faith because everything could be proved to be either true or false. By contrast, the world familiar to the rest of us is one in which faith assumes a major role in deciding truth. In particular, controversial hypotheses (such as the Darwinian and teleological theories about the origin of our species) are widely accepted as true, even though they have not been proved and perhaps never will be.

In 1931, however, the mathematician's fantasy world became like the more realistic world when the Viennese logician Kurt Gödel proved that there will always be mathematical truths that cannot be proved with logic. Suddenly, there was introduced into the mathematical world a formal role for subjectivity, since the only possible way of avowing an unprovable truth, mathematical or otherwise, is to accept it as an article of faith. (Subjectivity has always had an informal role insofar as mathematicians typically pursue hunches and other intuitive insights in formulating hypotheses.) It was an introduction whose consequences have not yet been fully seen: mathematicians have come to accept the possible existence of unprovable truths, but have not yet decided exactly what role—if any—faith should assume in their new, post-Gödelian world.

Actually, the mathematician's old world first began to give way to the new one around the turn of the century. Logic, which had always been perceived to be an infallible standard of proof, was even then under suspicion. Mathematicians such as the British logician Bertrand Russell came up with logical paradoxes that could not be resolved without some rehabilitation of traditional logic. To further complicate matters, mathematicians disagreed vigorously as to which of several possible remedies was called for.

These developments were particularly worrisome at the time because mathematicians were involved in a massive, collective effort to derive arithmetical truths in the same manner in which Euclid had derived his geometrical truths—that is, logically, from a small pool of starting assumptions. It was an important effort, because over the centuries arithmetic as a subject had developed haphazardly, and so much had been accepted along the way without proof that now no mathematician could be certain of its overall logicality. But the effort's success hung in the balance as the debate over how to repair logic continued into the 1930s.

Then, in 1931, Gödel quashed the effort and hushed the debate with his proof that there will always be mathematical verities that can't be proved with logic. This result was remarkable enough, but what made Gödel's achievement even more noteworthy is that he had used logic to incriminate logic.

The way he had done it was to consider an hypothesis

such as, Using logic, this hypothesis cannot be proved true. This is similar to the seductive question a man in a recent television commercial asks a woman: "Is it true that when you say no, what you really mean is yes?" Because of its peculiar self-referential wording, Gödel's hypothesis leads inescapably to a contradiction, whether one supposes it to be either true *or* false.

Imagine, on the one hand, that with logic we prove the hypothesis true. Its being proved true means that we have contradicted the assertion that "This hypothesis cannot be proved true," which means that the hypothesis is actually false. In short, by proving the hypothesis true, we have proved it false, which is nonsense.

Imagine, on the other hand, that with logic we prove the hypothesis false to begin with. This means that we have verified the assertion that "this hypothesis cannot be proved true," which means that the hypothesis is actually true. By proving the hypothesis false, we have proved it true, which, again, is nonsense.

At the same time, it is now evident as a result of this analysis that the hypothesis *is* true: "Using logic, this hypothesis cannot be proved true." We cannot say that we have proved the hypothesis true, because then we would just get involved in the logical contradiction all over again, but we *can* say that the hypothesis is, somehow, evidently true!

This was the revelation that led Gödel to surmise that there must be an indefinite number of mathematical hypotheses that behave this way, hypotheses that are extralogically evidently true, but that defy being proved true with logic. I will refer to these here as "unprovable verities."

Gödel's result does not specify just how many unprovable verities there might be in mathematics, nor does it clarify the nature of the extralogical faculty that would enable a mathematician to recognize an arithmetical truth that cannot be proved logically. In principle, therefore, mathematicians must now operate in a world where every mathematical hypothesis is potentially an unprovable verity, and where it is not yet clear what kind of extralogical principle should be brought to bear in evaluating the truth of an hypothesis that is suspected of being an unprovable verity.

One example in arithmetic of such an hypothesis is Goldbach's conjecture, which is that each and every even number can be written as the sum of two prime numbers. (A prime number is a number that can be cleanly divided only by itself and 1—1, 3, 5, 7, 11, 13, 17, 19, 23, and so on.) The even numbers 2, 4, 6, and 8, for example, can all be expressed as the sum of two primes (1+1, 1+3, 3+3, and 1+7, respectively).

This conjecture was stated in 1742 by the amateur German mathematician Christian Goldbach, who then took it to the eminent Swiss mathematician Leonhard Euler. Euler was unable to prove it, and—although the conjecture has been explicitly verified out to 2,000,000—no one else has been able to prove it yet.

In light of Gödel's result, this inability of mathematicians for almost 250 years to prove Goldbach's conjecture true suggests that it is probably either unprovably true or else just plain false, notwithstanding the current evidence in its favor. And as more time passes, the less certain it becomes whether the conjecture's resistance to proof signifies a greater likelihood that it is false or that it is unprovably true.

The increasingly urgent questions for mathematicians are how long they should wait and what criteria they should use to evaluate which of the possibilities is probably correct. The very use of the word "probably" in association with a decision about mathematical truth is alien to the pre-Gödelian fantasy world of mathematics, but now the importance of making such a decision, and an accurate one, cannot be understated.

If no decision is ever made one way or the other for want of an absolute proof, then a conjecture that might be true will exist uselessly in a kind of mathematical purgatory.

And if a wrong decision is made, then a conjecture that might be false will go on being used improperly as the basis for other arithmetical results. In any case, a decision would be a working, not a final, one.

The criteria question is especially momentous because Gödel's result warrants that the sole basis of any such criteria is of necessity some extralogical principle of faith. The eventual assimilation into mathematics of any principle of faith is bound to change the character of mathematics, although the exact nature of the change will depend largely on which specific principle is adopted.

The choice available to mathematicians among established principles of faith are wide-ranging, but two types represent the extremes. These are what I choose to call the secular and the mystical principles of faith.

According to the secular principle, an hypothesis is tentatively declared to be true if it is the simplest available explanation of the evidence. As such, it is a principle that reflects the conviction that an hypothesis should be intellectually esthetic—that is, rationally plausible and concise—as well as being an explanation of the evidence.

This is the principle of faith that in the sixteenth century guided Copernicus to believe in the heliocentric theory of the solar system and to disbelieve the more popular geocentric theory, even though both theories explained the available evidence. In order to reconcile the observed motions of the planets with the assumption that the earth occupied the center of the solar system, geocentrists were forced to advocate a rather complicated model of the solar system. In contrast to this, the orderly and symmetrical orbits imputed to the planets by the heliocentric model of Copernicus were intellectually more appealing. On the basis of that criterion alone, Copernicus preached heliocentrism as opposed to geocentrism.

This secular principle is now a canon of faith in modern science and is generally referred to as the principle of Occam's Razor (or the law of parsimony). It is named for the scholastic philosopher William of Occam, who in the early fourteenth century first put forward the idea that *non sunt multiplicanda entia preater necessitatem* (entities are not to be multiplied beyond necessity). As paraphrased in the nineteenth century by the celebrated Austrian scientist Ernst Mach, it instructs scientists to put their faith in only those hypotheses that explain the evidence objectively and concisely.

In contrast to the secular principle, the mystical principle of faith judges an hypothesis not only by how well it explains the evidence but also by whether it is consistent with a philosophy that assigns a purpose to everything. Also, whereas the secular principle implies that the truth of an hypothesis rests only on what we can verify with our five senses, the mystical principle implies that purposefulness is an additional, insensible factor that must be considered in evaluating the truth.

This principle of faith led the nineteenth-century teleologist archdeacon William Paley to believe in his hypothesis of divine creation called the "argument-from-design" and to disbelieve Darwin's hypothesis of evolution, even though both hypotheses explained the evidence available at the time. In order to explain the origin of species scientifically, Darwin was led to interpret the evident variety and harmony of the animal and plant kingdoms without imputing a purpose to them. To Darwin, the harmony and variety are the inevitable outcome of a nonpurposeful process that selects from among accidental genetic mutations only those that are best adapted to the environment. In contrast to Darwin's secular hypothesis, the mystical argument-from-design hypothesis explains the very same harmony and variety as being the handiwork of a purposeful agent. Unlike Darwin's hypothesis, the argument-from-design is true to the mystical principle of faith, and for that reason Paley was led to preach it.

In his famous debates with Darwin, Paley stated the argument-from-design something like this: Suppose that you're riding along in a train, when you notice some rocks on a hillside that spell out the message WELCOME TO MASSACHUSETTS. It probably would not occur to you to doubt that those rocks had been purposefully arranged, and yet it is conceivable that someone might take the position that the arrangement was a happenstance, effected over many, many years by natural geologic forces. This person might also hold a belief in the nonpurposeful appearance of human life on earth. However, no matter which way you might actually believe, if you depend on the arrangement of the rocks for evidence that you are indeed entering Massachusetts, you are necessarily conceding the truth of the teleological interpretation. Otherwise, your behavior would be irrational—you can't have it both ways!

Now consider the human eye. As with the rocks, the human eye as a sense organ can be interpreted either as having a purpose or as being a chance result of evolutionary forces. Whatever you believe, though, if you depend on your eyes for evidence of what the world looks like then you concede the truth of the teleological interpretation. Conclusion: since a similar argument can be made for each of our extraordinary human organs—including the brain—then it follows that the human body was purposefully constructed.

In the mathematical world there have always been two dramatically opposing philosophies from which it is possible that two distinct principles of faith will eventually emerge—principles that will be versions of the secular and mystical principles.

Adherents of the secularlike philosophy are called Formalists. They believe that mathematics is purely an invention of the human mind, and, according to them, mathematical hypotheses do not refer to anything real. Proving an hypothesis only means that it is a successful invention, much like an airplane that actually flies.

In opposition to the Formalists are the Platonists, adherents of a mystical-like philosophy. The Platonists believe that mathematics, like science, is an infallible means of discovering truths that exist independently of the human mind. For them, mathematical hypotheses do refer to real things—to disembodied mathematical truths—and proving one is tantamount to verifying a scientific hypothesis such as geocentrism.

I can foresee the possibility of Platonists and Formalists squaring off in a debate over how to evaluate the truth of longstanding hypotheses, such as Goldbach's conjecture, whose truth or falsity might not be decidable with logic alone.

The formalists, on the one hand, are liable to look upon such hypotheses as unsuccessful inventions. They might be inclined to believe that if an invention has not been proved—hasn't "flown," as it were—after 250 years of effort, it is probably never going to fly.

The Platonists, on the other hand, of which Gödel was one, are likely to regard mathematical hypotheses such as Goldbach's conjecture as they would a scientific hypothesis such as modern Darwinism—impossible to prove, maybe, but commended to us nonetheless by a growing body of evidence.

Given that Gödel's result guarantees that logically unsettled, and unsettleable, hypotheses will always exist, it is conceivable that disagreements such as this will become as common in the mathematical world as disagreements over the existence of God are in the real world.

In the meantime, the world of today's mathematician is one not only in which truth is not synonymous with logical proof but also in which merely trusting in the validity of a logical proof is itself a matter of faith. This is because Gödel not only showed that any logical system is unable to prove all the mathematical statements that are actually true, but also that any system of logic is unable to prove its *own* logical consistency. Believing in logic, in other words, is no less subjective a frame of mind than believing in, say, a secular or mystical principle of faith, because even logic itself cannot be verified logically or objectively.

This is not to say that the mathematician's traditional

perception of logic as something not subject to doubt is necessarily irreconcilable with Gödel's result. Consider geocentrism, which, like logic, was once widely perceived to be indubitable on the basis of a mystical-like belief in the divine significance of life on earth. In recent years, geocentrism has been reconciled with the realities of modern science by an awareness that earth does occupy a unique place in the solar system—the only one conducive to life—even though that place happens not to be the geometrical center. Perhaps in coming years, mathematicians' long-standing belief in the infallibility of logic will be reconciled in a similar fashion with the realities of modern mathematics. Gödel's result displaced logic from the center of the mathematician's world, but in so doing it has challenged us to locate a noncentral but unique place for that rarity among principles of faith: a passionate form of belief that avows its own shortcomings.

Related Essay

A Certain Treasure

PART THREE

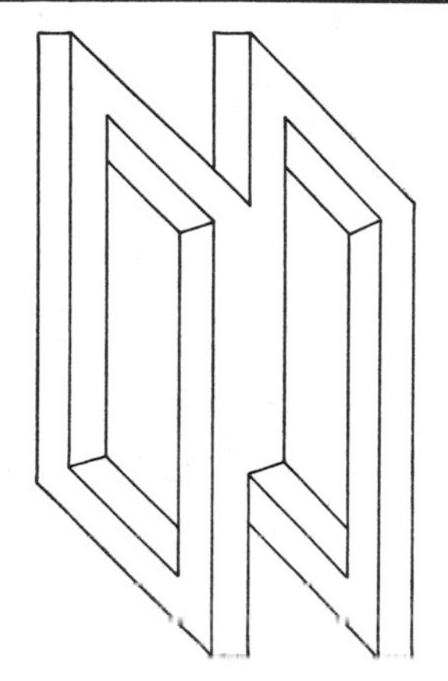

OPTIMIZING

Probability Theory and Statistics

THE CLOUDY CRYSTAL BALL

> In the small number of things we are able to know with any certainty . . . the principal means of ascertaining truth are based on probabilities.
>
> Pierre Simon de Laplace

Everything that physicists have learned about the natural world invites us to accept uncertainty. Even the behavior of atoms, which scientists once believed to be completely predictable, is now known not to be so.

To the extent that we can predict unpredictable behavior, we depend for doing so on the mathematical theory of probability and its modern offshoot, statistics. They are, in some cases, the only way we have of making our unpredictable world seem predictable.

On a scale of unpredictability, behavior that can at best be described in terms of probabilities, such as the behavior of dice, lies somewhere between totally predictable behavior (the motion of the moon about the earth, for example) and completely unpredictable behavior (the output of a random number generator).

At the one extreme, completely predictable or deterministic behavior is most commonly seen in objects whose comings and goings are entirely dictated by rigid laws. The entire future of such an object can be predicted from a knowledge of its present state of being and of the laws that govern its evolution. For instance, all astronomical bodies behave deterministically because their motions are strictly choreographed by gravity.

It was because of the success early on in predicting the motion of such heavenly objects as Halley's Comet that many scientists of the eighteenth century assumed that they could eventually predict the behavior of *anything*, including a human being, with equal precision. Needless to say, we are still far better at predicting the return of Halley's Comet (due next in April 1986) than we are at predicting almost everything else.

At the other end of the scale, completely unpredictable behavior would be expected from entities whose conduct is not governed by the dictates of any law. Such arrantly capricious behavior, it turns out, is quite rare and perhaps even nonexistent: Scientists in both physical and social disciplines have yet to discover a natural phenomenon that is wholly unpredictable, or chaotic. This seems to suit most scientists just fine, since it is an article of faith in science that the workings of the universe are entirely rational; unpredictable behavior is anything *but* rational.

The absence of any natural models of chaos does, however, make it a difficult challenge for mathematicians and scientists interested in studying chaos to create models that are perfectly random. To my knowledge, in fact, the behavior of every model of chaos devised so far has turned out to have some predictable pattern to it.

Between the two extremes of perfect predictability and perfect unpredictability is the kind of behavior we see most often in the world. It is behavior that, while not governed by any law at the individual level, *is* governed by strict laws at the collective level. Such is the paradoxical nature of probabilistic behavior (that is, behavior that can only be described in terms of probabilities).

For instance, in a roomful of air molecules, which is a system known to behave probabilistically, the speed of any single molecule can be anything from zero to 186,000 miles per second (the speed of light). Speaking anthropomorphically, any single molecule has a free choice in selecting the speed at which it prefers to travel. Simultaneously, however, there is a law of thermodynamics (the physics of energy) that imposes an overall predictable pattern on the collection of "free choices." According to this law, some speeds must be represented more or less than others, depending on the temperature of the room.

The situation in the roomful of molecules would be roughly analogous to that in a roomful of hungry people after an announcement has been made that each person can have a choice of three lunches, but that there are only a certain number of lunches of each kind available. Such a system is at once chaotic and deterministic, and it can seem predictable even though it is not.

In explicating the mathematics of such conduct, the theory of probability provides us with the mathematical means to predict the future of probabilistic systems with as much precision as possible. For the roomful of air molecules, this means that we can predict its evolution as a whole; in this case, its future is simply that it will remain confined to the room at a fixed temperature unless otherwise tampered with. The specific future of each individual molecule is not part of the prediction.

First and foremost, the theory of probability clarifies what it is exactly that we mean by the term "probability." It is defined in hypothetical terms, as the number of times out of a hundred some particular outcome would be realized if a probabilistic system such as a pair of dice was allowed to fall into place randomly an infinite number of times. Saying that the probability of throwing a pair of sixes is about three out of a hundred, for instance, means that in the infinitely long run, a pair of sixes will come up about three times out of every hundred tosses; it does *not* mean that three pairs of sixes will come up each time we toss the dice a hundred times.

Clearly, the definition of probability is impossible to apply if it is taken literally. We cannot actually toss a pair of dice, or test any other probabilistic system, an infinite number of times; therefore, we can never measure probability as it is theoretically defined. However, according to the theory of probability, we *can* measure something very close to the theoretical probability by observing the outcomes of a probabilistic system a very large number of times. The larger the number of times this is done, the closer a measured probability will zero in on some specific value. This value is—or is very close to—the theoretical probability.

Observers who have kept track of human births and coin flips for decades report that the probability that a baby will be a boy is about the same as the probability that a coin will come up tails—something around 51 out of 100. These are measured probabilities only, but they are based on such a large number of "tosses"—tens of millions, in each case—that we can expect they are very close to being the precise theoretical probabilities.

If the theory of probability clarifies what we mean by probabilistic behavior, then its offshoot, statistics, gives us the know-how for recognizing the deterministic features in probabilistic behavior. (These are the features, you'll recall, that impart to probabilistic behavior what predictability it does have.) Thus, with their techniques—or "black arts," as one colleague of mine puts it—statisticians are able to forecast the future fates of behavior that often seems quite unpredictable. In that respect, statisticians are the nearest thing to soothsayers that modern mathematics offers.

Central to the statisticians' efforts is the notion of statistical sampling. That is, everything that a statistician concludes about the probabilistic behavior of a population, whether it's a roomful of molecules or a countryful of people, is gleaned from studying the behavior of a sampling of that population. In principle, the individuals included in a statistical sample should be representative of the population in every respect. In practice, it is usually only feasible to select a sample that is representative in some of the more obvious ways. In a population of people, these ways might include race, sex, age, and tax bracket.

Statisticians usually have their own personal techniques for selecting representative samples; how successful each is in this single respect naturally determines how accurate their information will be about the population as a whole. The techniques vary, according to the efficiency with which a sample is collected, the way in which it is verified as representative, and other factors that depend on the specific technique that is used.

George Gallup is one statistician whose technique—a trade secret—is widely acknowledged to lead to especially reliable predictions about probabilistic behavior. Evidently, Gallup uses a methodology that is especially quick (his people use telephones and door-to-door interviews to collect their information) and reliable (his organization has an extensive library of demographic information about the American people; such information guides the statistician in assembling a statistical sample that is truly a microcosm of the population).

Once a sample is collected, statisticians evaluate the behavior of its constituents for any deterministic features that will enable them to make predictions about the behavior of the population as a whole. These features are called statistics, and among the growing variety that statisticians have come to identify, three play a particularly large role in predicting the outcomes of probabilistic behavior: statistical percentages, statistical averages, and statistical correlations.

A statistical percentage is simply the percentage of individuals in a sample that behave in a particular way. In the case of air molecules, this statistic might refer to the percentage of molecules that have large speeds. Knowing such a percentage could help the statistician to predict just how rapidly the molecules would diffuse if they were suddenly liberated. Part of the challenge for the statistician is in deciding which percentages are useful to know and what exactly they predict once they are known. A question of this nature has no single answer; inexactness is one reason that statistics is commonly referred to by mathematicians as an art as well as a science.

The A. C. Nielsen organization has made an industry out of measuring statistical percentages for the information of television executives. Their statistical sample is constituted by a mere 1500 television sets located in households of people who have been determined to be a representative cross-section of Americans. The members of each household are requested to keep a log of their viewing preferences, and on the basis of this information, the Nielsen organization figures out the percentages of people in their sample who watched the various shows. These percentages are called the Nielsen share ratings, and they are used by television executives to decide—and can be used by the rest of us to predict—which shows will be coming back for another season.

Our viewing habits have become so well-known to the television industry that in many cases it can be predicted which shows will do well even before they have aired. This again illustrates the paradoxical nature of probabilistic behavior—at the individual level we are unpredictable, in that each of us has an ostensibly free choice in which television show to watch; but at the collective level we are entirely predictable.

The statistical average is another feature of a statistical sample that betrays something of the predictability of probabilistic behavior. It most commonly refers to average *behavior*, which is the behavior that is representative, in some sense, of the various behaviors observed in the sample; these in turn are representative of the myriad behaviors of the actual population. The more prevalent a particular behavioral trait is in a population—a preference for chocolate, for instance—the more weight that trait will have in determining the average behavior.

Insurance companies are especially known to rely on the notion of statistical average and on statistics in general. Statistical averages are the basis of their actuarial tables, which are the basis of their profits. The fact that life insurance companies have been making profits for years is evidence enough of the predictability of probabilistic behavior, and of the success with which mathematics can be used to forecast something as apparently unpredictable as when we will die. Today, life insurance companies can tell us very precisely the average longevity of people with our particular habits, personality, genetic heritage, race, income, and so on.

The third statistic commonly used to reveal the deterministic aspects of probabilistic behavior is the statistical correlation. In order to surmise a correlation, the statistician must work with at least two different samples. One of them is always a sample that is chosen to represent the population being studied; this is called the control sample. The other is a sample that is predominantly or entirely constituted of individuals who are distinguished by some specific trait, such as cigarette smoking. If a thorough comparison of the two groups reveals any notable disparities, these are seen as linked to the one trait (smoking) that is known to distinguish the samples. Actually, in the case of smokers versus the control group, scientists determined that there was a smaller percentage of lung cancer victims in the control sample than in the sample of smokers. It can therefore be said that lung cancer is correlated with smoking.

Such a correlation is used to predict that someone who smokes is more liable to die of lung cancer and to die at an earlier age than someone who does not smoke. Correlations of this kind are of particular interest to life insurance companies because they are indicators as to which clients are good risks and which are bad risks. Some insurance companies have used statistical correlations as the basis of advertising campaigns offering special low-premium policies to low-risk clients. For example, we now see companies offering dramatically reduced life insurance rates to healthy, nonsmoking, middle-aged white women who don't drive; this is a consequence of the high correlations that have been found between lung cancer and cigarette smoking; highway deaths and young adulthood; violent death and young black men; and a lower life expectancy for males generally.

Regrettably, the same means that we use to predict the outcome of probabilistic behavior can be used to *influence* it, either unwittingly or deliberately. By informing us about our predictable collective behavior, for instance, statistics can influence free choice at the individual level to such an extent that it nearly eliminates the chaotic, unpredictable aspect of untampered-with probabilistic behavior. Under such an influence, our behavior could become as deterministic as the motion of Halley's Comet.

A concern was widely expressed about the possibility of this having happened to some modest but notable degree in the 1980 presidential election between Ronald Reagan and Jimmy Carter. Hours before the polls were scheduled to close on the West Coast, the television networks were already predicting, on the basis of their samplings of East Coast results, that Reagan would win the election. Although we don't know that they actually did influence West Coast voters, the predictions might have done so by discouraging or encouraging voters to cast their ballots for the leader or the underdog—or not to vote at all.

Statistics can also be misused to discriminate against certain behavior and, by association, certain people. A perception of some statistical average, for instance, is the basis of an employer's reluctance to hire someone who does not match the profile of the efficient company person. In a similar way, restaurant owners often do not cater to single women, because statisticians have determined that the average single woman tips less and monopolizes a table longer than the average anyone else.

Even when statistics *is* applied properly and with sensitivity, it is not clear how useful it can be in predicting those outcomes of human behavior that have historically proved to be the most significant influences on the future: statistics can only reveal to us the predictable aspect of historical human behavior, not the unpredictable one, and yet it is the unpredictable outcomes of human behavior that have historically proved to be the most consequential.

Innovations such as the steam engine and the computer came from the minds of persons who were anything but statistically average. And in general, the coming into the world of people such as Jesus, Socrates, Newton, and Einstein is so unpredictable, and also so influential, that the effect of their existence on our historical behavior is almost totally chaotic.

It is natural to wonder whether we shall ever find a mathematical means to predict the chaotic aspect of probabilistic behavior, just as we have found a means in statistics to predict the deterministic aspect of it. If we ever do discover such a means, then human behavior will be made to seem wholly deterministic, and we shall be able to foresee the entire future of our species and our universe. The probability of this ever happening might seem small to us now, but it is not zero. For precisely because one aspect of our behavior *is* chaotic, *is* unpredictable, the seemingly impossible will always be probable.

Two-Person Game Theory

BETWEEN CHECKERS AND CHESS

If but a beam of sober Reason play,
Lo, Fancy's fairy frost-work melts away!

Rogers, *Pleasures of Memory*

It has often been said that humans can be distinguished from the other animals by their ability to reason. This has always struck me as little more than a rather ambiguous way of stating our conceit as a species.

What actually constitutes an "ability to reason," how can we judge it, and how much of this ill-defined ability do we have? The answers to these questions might enable us to better evaluate our relative standing among animal species and put us in our rightful place, once and for all.

One means that I will propose here for accomplishing this involves game theory, the mathematical study of rational conflict. My purpose will be carried out by focusing on two-person zero-sum game theory, which is the mathematical study of rational conflict between two persons in which one person's win is necessarily the other person's loss—as it is, say, in a game of tick-tack-toe.

The main reason that this type of game theory is an appropriate means for evaluating an ability to reason is that it is concerned with contests that are strictly rational; emotions play no role in two-person zero-sum game theory. (Emotions are a factor only in two-person nonzero-sum games, in which players' losses and gains need not always balance out—as with business competition in an expanding market; for example—and in games involving three or more persons.) The fact that success is determined strictly by a player's ability to reason makes two-person zero-sum games the perfect theaters for playing out our "species evaluation program."

In game theory, certain specific assumptions are made about the nature of a contest in order to ensure its rationality; these assumptions are, in a sense, the rules of the house.

First and foremost, it is assumed that both players are equally competent—that is, they will match wits using their matched wits. Players are expected to deliberate all their options before deciding, dispassionately, to make a particular move. And, for our purposes, we will also assume that players are equally and completely enlightened at all times about all aspects of the game's progress. Games in which this assumption is enforced are called games of perfect information. In such games, including checkers and chess, there are no secrets between players, as there are in games of imperfect information, such as poker and bridge.

The most important result to date in game theory about games of this kind will form the basis of the species evaluation scheme presented here. This result, proved in 1921 by the German mathematician John von Neumann, is that every conceivable two-person zero-sum game of perfect information has what is called an optimal strategy uniquely associated with it. The optimal strategy takes its name from the fact that each player in a two-person zero-sum game of perfect information can do no better than to follow his or her role in the optimal strategy for that game.

In order to understand what "can do no better" means,

recall first that a player in game theory is always up against an opponent that is his or her equal in all relevant ways. Aware of this, "Player A" knows that throughout the game "Player B" is striving with equal competence and tenacity to accumulate gains. Under these circumstances, Player A's only rational strategy is to minimize Player B's success at accumulating gains and to thereby minimize his or her own losses. But as this same reasoning is of course also going through the mind of Player B, the best that either player can hope for is that the game will end in such a way that he or she is left in the best of the worst possible situation. This is the definition of optimal strategy. By following it, a player is left content and relieved, thinking something like: Considering that the other player was my intellectual equal and had my same means and aspirations to win, I could have ended up worse off, but no better off.

An optimal strategy will always fall into one of two categories, fair and unfair, depending on whether it guarantees the players a tie game or a predictable win for one of them. Tick-tack-toe, which is a two-person zero-sum game of perfect information, has an optimal strategy that is fair; there is a way of playing tick-tack-toe that guarantees one a tie against any opponent. (The first moves of the optimal strategy for tick-tack-toe are decisive: the first player marks a corner position and the second player responds by marking the center position.)

In Nim, an ancient two-person zero-sum game of perfect information, each player in turn removes any number of sticks from a pile, and the winner is the player who picks up the last stick. Nim has an optimal strategy that is unfair because by following it, the first player will always win, and the second player will always lose.

The existence of optimal strategies for two-person zero-sum games of perfect information offers us an opportunity to gauge the reasoning ability of players of these games. The opportunity lies with the observation that a game loses its challenge, and appeal, for players who have the ability to reason out its optimal strategy—that is, if both players know a game's optimal strategy, they will know the game's outcome even before they start playing. Thus, they would have no incentive to play the game at all.

In the case that the game's optimal strategy was fair, players who have figured out the strategy will know that their play will certainly end in a tie. In the case that the game's optimal strategy was unfair, players who have figured out the strategy will know ahead of time who will win and who will lose. In both cases, the game becomes a pointless predetermined exercise for players who have reasoning ability enough to figure out its optimal strategy.

We can get a pretty accurate gauge of the reasoning capacity of a species, therefore, just by observing which games its adult members play. In other words, we can assume that if the adults of a species play tick-tack-toe, it is because they find it challenging. This implies that they have not figured out the game's optimal strategy. Their playing tick-tack-toe, therefore, would place their ability to reason somewhere below that of another species for which tick-tack-toe might be recognized to be a predetermined exercise and disdained as mere child's play.

Applied to human beings, this evaluation scheme immediately establishes us as a trans–tick-tack-toe species. Our adult ability to reason is hardly challenged at all by tick-tack-toe, which is why the game is popular mostly with young children who have yet to figure out its optimal strategy.

We come close to the limit of our capacity to reason with the game of checkers. Many experienced checkers players have been able to work out a kind of working strategy, which makes it entirely plausible that it is within our reach to reason out the one and only true optimal strategy that according to game theory does exist.

A game that comes closer still to evidencing the limits

of our reasoning abilities is chess. Chess tends to be a more popular game than checkers among intellectuals, which might be an indication of its greater challenge to our minds. People are more inclined to be impressed by a master chess player than by a master checkers player. Chess masters in Russia are nurtured and treated like royalty. And even though in this country we do not lavish such hospitality on even our best chess players, we are given to thinking of them as geniuses of some kind. Bobby Fisher is a recent example of an American chess player whose talents, and eccentricities, attracted a prodigious amount of notoriety.

Another indication of the enormous and exciting challenge of chess for us is that of late we have even enlisted the aid of computers to figure out the game. And yet even with such assistance, although it has helped us to formulate some successful chess strategies, we still have not been able to reason out the unique optimal strategy.

Chess would not command such effusive esteem and attention from a species whose reasoning capacity had enabled it to work out the game's optimal strategy. For a trans-chess species, chess would be as trivial a distraction as tick-tack-toe is for us. Trans-chess players would know ahead of time which specific moves to follow out of the possible twenty trillion, trillion, trillion, trillion, trillion moves in order to guarantee themselves a predictable outcome—which would either be a tie or a predictable win for one of the players, depending on whether the optimal strategy for chess is fair or unfair (which of these it is, we do not know). The only way in which chess would be any fun at all for players of a trans-chess species is if they agreed not to play by the optimal strategy.

We could get a very specific idea of where we stand along our theoretical scale of reasoning ability by taking a particularly thorough inventory of which games challenge us and which ones do not. But on the basis of the few observations made, it can be estimated that the limits of our capacity to reason as a species places us somewhere between the limits of checkers and chess.

In putting us in our proper place, as it were, game theory used in this fashion reminds us that we do have measurable limits. In particular, it is a pointed reminder to those of us who would use definitions of the human species as a way of exalting it over all other species. Being informed of our place between checkers and chess reminds us that an ability to reason may be possessed in degrees, and that on the same scale, other species might just as conceivably rank above us as below.

Game theory applied to evaluate our capacity to reason also tells us something else about ourselves: We are distinguished by our playfulness as much as we are by our reasoning level. Our fascination with games is evidence not of our ability to reason, as one might guess, but of our lack of a greater reasoning capacity. The higher a species falls on the scale of reason, the less playful its members can be expected to be. By recognizing this aspect of what game theory tell us about what it means to be human, we should all be happy indeed that we are not at the top of the heap!

Related Essay

The Call of the Wild

Three-Person Game Theory

THE CALL OF THE WILD

> The adventurer is within us, and he contests for our favor with the social man we are obliged to be.
>
> William Bolitho

Economically hard times seem to bring out both the best and the worst in people. All at once, they become more covetous and more generous, competitive and cooperative, pragmatic and philosophical, anxious and resigned. Perhaps people cannot decide how to behave because they cannot decide whether their long-term interests are better served by behaving individualistically or socially.

In game theory—the theory of rational conflict—this tension between a person's private and social persona is seen to be the unavoidable consequence of cooperation. People may sometimes feel that they could do better if they handled things individualistically, without the demands made on them by society. Game theory predicts that there will be natural difficulties associated with cooperation, but it also indicates that the alternative would be worse.

Human beings are social animals by circumstance perhaps as much as by nature. Incentives to cooperate, to form a society, can present themselves even in situations involving only two persons.

Consider, for instance, the two-person game known as the prisoner's dilemma, in which a district attorney wishes to extract a confession from either of two men suspected of committing a murder together. The D.A. makes the following offer: "The one who turns state's evidence will go free, and the other will probably get twenty years. However, if both of you confess, then each of you will receive a five-year sentence. And if neither of you confesses, then each of you will get a one-year sentence anyway, on a lesser charge." The dilemma exists in that if each prisoner privately opts to confess, reasoning that at worst he will get five years and at best he'll go free, then both surely will end up with the five years.

An incentive for less individualistic behavior exists here, because if the prisoners agree to trust each other not to confess, then the worst each will receive is a one-year jail term. At the same time, such cooperation raises possible problems. For instance, each of the prisoners must consider the possibility of a double cross. Furthermore, cooperation sometimes not only complicates a game, it actually makes it irresolvable.

This is the case with the apparently simple game of three people deciding how to divide $300. It is assumed that the majority rules in all decisions. If there was total voluntary cooperation, each person would probably receive $100. But if we impute to these players a realistic modicum of selfishness and opportunism, we can see that they would realize that two of them could cooperate against the third and receive $150 apiece. The outcast could still get back into the game by trying to lure one of the conspirators into a new alliance, promising him the *larger* share of $300 split between them. But even if this offer resulted in a tentative agreement, the jilted conspirator could come back with a counteroffer of the same kind, and so on indefinitely.

It is the inevitability of this kind of conflict in any game involving three or more persons that makes them mathematically irresolvable. That is, if three or more players are left to compete and cooperate unrestrictedly, we can expect that their "society" will be unsettled.

No doubt the English philosopher Thomas Hobbes would have embraced this observation as evidence for his thesis that humans outside of society will behave anarchistically. Three people vying for a share of $300 might not constitute an anarchy, but clearly they are in want of something that will render their game resolvable. The English philosopher John Locke also believed that humankind's natural state was an anarchy, a state in which each person behaves individualistically. In his *Second Treatise of Civil Government,* he wrote: "To avoid this state of war . . . is one great reason of men's putting themselves into society and quitting the state of nature." Thus society has a man-made purpose, but in order to exploit it, a person has to put "himself under an obligation to every one of that society."

Historically, we have agreed on the desirability of "quitting the state of nature" far more than we have on what is the best way to go about it. According to the theory of games in which there is cooperative behavior, a stable society is one that is equitable to its members.

This can be interpreted in many different ways, however. Even if a group of people agreed, for example, to restrict the interpretation of the word "equitable" to only financial terms, there would probably be no agreement.

One person, calling himself a socialist, would eliminate competition and have everyone agree ahead of time to split their collective resources evenly. Another person, calling himself a capitalist, would retain competition and only require that it conform to certain rules of fairness, so that each person's compensation would be commensurate with his enterprise. And then there would be those who would disagree as to the meaning of equitable as applied to civil liberties. One person, perhaps a fan of Ayn Rand's, would have the society's government solely concerned with "the protection of individual rights." Another person, in agreement with John Stuart Mill, would have the government provide for the "common good."

And never mind philosophical disparities; mere personality differences could sabotage the prospects of a group's becoming a cohesive society. In terms of game theory, the stability of any coalition depends decisively on the payoffs to its members. It is generally accepted that a person will wish to join or remain in a coalition only if his payoff for doing so exceeds the payoff he expects to receive by going it alone. Clearly, then, someone with confidence in his ability to compete, perhaps because of his success in the state of nature, would be pulling for a capitalistic system, and someone who was naturally diffident and deferential would just as likely be opting for a socialistic system.

Obviously, many societies have managed to form in spite of these obstacles. And perhaps an even more formidable challenge than its struggle to come into being is a society's struggle to persist. Charter members eventually die off; their descendants, not having lived in the state of nature, may not have the same allegiance to the coalition and may not justify its existence with the same enthusiasm.

In capitalist societies such as the United States, one of the ongoing questions is just how much freedom should be allowed an enterprising member in his efforts to accumulate personal wealth before his efforts become detrimental to others. How far can such personal freedoms go before they return the society to as primitive a situation as that of three people splitting $300?

In his classic work *Wealth of Nations,* Adam Smith stood firmly on the side of a laissez-faire policy, believing that a society enhances its wealth by abiding the relatively unrestricted freedom of every person to enhance his own wealth.

"As every individual . . . endeavors as much as he can . . . to employ his capital in support of domestic industry . . . , every individual necessarily labours to render the annual revenue of the society as great as he can."

Smith's idea is remarkable if only because it seems to suggest that up to a point, the more individualistically people are allowed to behave, the more socially they will behave, albeit inadvertently. But it also suggests that a society in which people are coerced by legal regulations to cooperate on its behalf will not be as prosperous as one in which people cooperate only if it suits their individual purposes to do so.

In 1951, the American mathematician John Nash published a valuable extension of the existing theory of cooperative games in a work appropriately entitled *Non-Cooperative Games*. The theory seems an especially valid means of treating the concept of laissez-faire because it concerns many-person games in which each player (1) strives to maximize his individual wealth, (2) is bound to the other players only by a few common rules, and (3) can communicate any wish to cooperate only by the way in which he behaves and not by direct communication. Though Nash's premises are consistent with those of Smith's thesis, his conclusions do not fully corroborate Smith's stated expectations.

The first thing Nash discovered about the repeated play of a noncooperative game is that the outcomes inevitably evolve toward one of two kinds of equilibriums, which he characterized as optimal and suboptimal. What drives this tendency is the fact that only at these equilibriums is a player in the most likely position to be either in line for a reward or immune against paying a penalty. But although either kind of equilibrium is advantageous from an individual's viewpoint, Nash found that only the optimal one is desirable in terms of the group's collective interests.

Nash's results can be illustrated by way of the noncooperative game called stagnation, in which each person has the option of either staying with or breaking away from the group. The rules, which are meant to depict realistic group psychology, specify that if a majority breaks away, then each person in that majority will receive one dollar. If, however, a minority chooses to break away, then each maverick will be docked ten cents, while those in the majority staying behind will be neither penalized nor rewarded. Players cannot explicitly collaborate before making their moves, though after playing the game several times they may be able to second-guess one another's likely actions and attempt to communicate indirectly through repeating a certain pattern of behavior. Laissez-faire or not, in a large society where direct communication is not always feasible or practical, Nash's particular assumption seems valid enough.

This game happens to have an optimal equilibrium in which all players opt to break away, and each therefore receives a dollar. It also has a suboptimal equilibrium in which all opt to stay with the group and thus avoid the risk of paying a maverick's penalty. Once the game-playing reaches either of these equilibriums, there is virtually no incentive for any individual player to budge from his position, since to do so would almost certainly result in a personal loss. A policy of laissez-faire apparently promotes a kind of pack mentality among players; in trying to anticipate one another's moves for the purpose of enhancing their individual well-being, they can lead the entire group to its least desirable fate as easily as to its most desirable one.

It is not difficult to recognize the suboptimal equilibrium of the stagnation game, in its various guises, in many of today's societies. In an ailing economy, major banks may aggravate the overall situation by charging high interest rates, partly because each is fearful of risking a loss by being the first to lower them. Passengers waiting to board a plane on a first-come, first-served basis will suddenly stampede when, nearing the time of departure, they notice one person inching his way toward the gate. More seriously, in a society infested with crime, a citizen may choose not to get involved to help prevent a crime that is happening before his very eyes for fear of inviting personal injury. In these examples, a person's "every man for himself" reaction to the group situation merely exacerbates the problem and drives the society deeper into its suboptimal equilibrium.

To prod a group out of a suboptimal equilibrium generally requires that a penalty be attached to a decision to stay with the group. Better yet, an incentive might be tied to any decision to break away—for instance, positive reinforcement in the style suggested by the psychologist B. F. Skinner. Both options seem to make it inescapable that some central authority is needed to coordinate either a voluntary or legislated effort to avoid or escape from a suboptimality.

In the first of my examples, the U.S. Federal Reserve Bank or some comparable agency could try to trigger lower interest rates by increasing the money flow. In the second example, an airline company might require passengers arriving at the terminal to take a numbered ticket that would determine their positions in a boarding queue. In the final example, laws might be passed that would punish "bad Samaritans," or, more realistically, a local government might promote the founding of neighborhood watch programs.

Urged upon each of us who belongs to a society that finds itself in a suboptimal slump is the challenge to initiate action that might stimulate his or her society into moving toward its optimal equilibrium, even though this may be at the risk of personal loss. Enduring economic woes, rampant crime, high unemployment rates—crises capable of disassembling societies—are perhaps the closest approximation to a state of nature that modern humankind experiences. In the midst of these crises, each of us is faced with something of the same choice our ancestors once contemplated before they finally opted to band together. The many-person game that is society thus boils down to the two-person game that is within each of us. It is a hoary and not altogether rational contest between our social and antisocial tendencies. Today, as in times past, the central choice still lies between the beckoning of society and the call of the wild.

Related Essay

Between Checkers and Chess

Topology

A STRETCH OF THE IMAGINATION

For what is form, or what is face,
But the soul's index, or its case?

Nathaniel Cotton, *Pleasure*

Although we can expect that the normal aging process will transfigure our appearances somewhat, we can also expect that it will leave much about us unaltered. I was reminded of this at a recent reunion of my college classmates. Despite their older appearances, each was readily recognizable to me by a unique mannerism or personality. Within a short time of my arrival, I had practically lost sight of the older appearances—in some ways it was as though nothing had changed.

Others have found similar pleasure in discovering the existence of permanence in the midst of change: in philosophy, the Eleatics; in religion, the Buddhists; in painting, the cubists; and in mathematics, the topologists. Actually, the concept of permanence in change is fundamental in several other areas of mathematics as well. It is at the heart of group theory, which is the study of symmetries. In the 1870s, the concept of permanence in change was used by mathematician Felix Klein to classify the various known kinds of geometry, including topology.

Topology itself is the geometry that is concerned with those properties of a thing that are not destroyed through bending, stretching, and twisting, the three specific elements of a topological transformation. For instance, any three points along the circumference of a hoop will keep their relative positions no matter how much the hoop bends, stretches, or twists; the middle point will remain in the center, and the outer ones will remain on either side of it. A property such as this is called a topological invariant, because it survives the rigors of a topological transformation.

In its concentration on the immutable qualitative properties of things, topology complements the more widely familiar metric geometry, which is concerned with the precise measurements of an object's angles, width, breadth, and so on. This is the geometry that we all studied in school and that was originally developed by the Egyptians more than 2500 years ago for surveying and architectural purposes; in fact, the word *geometry* comes from the Greek roots *geo* and *metrein,* which together mean literally "measurement of the earth."

To the metric geometer, for whom the size of things is important, a widening circle represents change. To the topologists, for whom topological invariants are what matter, a widening circle represents constancy. The metric geometer calls our attention to the circle's increasing circumference. But the topologist ignores the superficial changes in the circle's appearance and notes that throughout the widening process, the circle remains a simple closed curve with an unambiguous inside and outside.

In a sense, then, if metric geometers study the transitory aspects of geometrical objects, then topologists study their souls. An object's "soul" is the collection of its topologically invariant properties, because they, more than anything else about it, are the essence of the object. What is a circle, for instance, if not a closed curve with an inside and outside? And what is a doughnut if not a closed surface with a hole in it? By focusing on what it is about things that endure even as everything else about them changes or is susceptible of changing, topology studies the most fundamental aspects of geometrical existence. Without topology, geometry would be as incomplete as philosophy would be without metaphysics.

In the time since 1736, when the Swiss mathematician Leonhard Euler founded their subject, topologists have learned a great deal about the constitution of and relationships between the souls of geometrical objects. They have learned, for instance, that every object has a soul (that is, every object is characterized by some collection of topological invariants) and that, often as not, the only way of recognizing that two very different looking objects are actually related is by seeing that their souls are identical. But perhaps the most significant outcome of topology is the specific identification of topological invariants. In knowing them, topologists know the essential features of geometrical existence.

Three of the most basic invariants identified to date by topologists are an object's dimension, number of edges, and number of sides. Each of these properties of an object remains unaltered even though the object may be twisted, bent, or stretched.

An idealized strip of paper has two dimensions (an idealized surface has no thickness), one edge, and two sides. These properties are its soul, because, disfigure the strip of paper as we will, it will *always* have two dimensions, one edge, and two sides. The same is true for the idealized surface of a basketball (two dimensions, no edges, and two sides), the idealized surface of a cylinder (two dimensions, two edges, and two sides), and every other conceivable object.

Topologically speaking, kindred souls are called homeomorphs. These are objects that by virtue of having identical topological invariants are considered topologically equivalent. A deflated basketball, misshapen as it is, is homeomorphic to a fully inflated one, because both have surfaces characterized by the same invariants. In fact, any of the shapes that a basketball could take on by being bent, twisted, and stretched are homeomorphic.

The concept of homeomorphism implies that the topologist's view of geometry resembles some artists' view of reality. Some artists who look upon an object tend to see all the different ways they might depict the object and still communicate its essential being. Monet must have reacted this way to the Rouen cathedral in Paris before he decided to paint it at many different times of the day. Similarly, topologists looking upon an object are liable to see in it all the different, homeomorphic shapes that it might have and still be essentially the same object.

In addition to unifying things in geometry, the notion of homeomorphism serves to categorize them. For instance, the homeomorphic variations of the basketball constitute a category of kindred souls that includes all conceivable objects with two dimensions, no edges, and two sides. This and every other category is strict in that it is impossible for an object in one category to be topologically transformed into an object in another category. It is impossible, because the topological invariant, the soul, of the one object can never be bent, twisted or stretched into the topological invariant of the other object. Topological categories are strict, in other words, just because an object is categorized according to the nature of its soul, and its soul is indestructible against topological transformations.

Against nontopological transformations, however, souls are quite transmutable. With glue and scissors we can accomplish what bending, twisting, and stretching cannot, namely, changing the topological invariants of one object into those of another object. A strip of paper whose ends are glued together, for instance, becomes a cylinder. Thus, with a bit of glue, we've changed a two-dimensional, one-edged, two-sided soul into a two-dimensional, *two*-edged, two-sided one.

If we give one end of the strip of paper a half twist before doing the gluing, we can create an even stranger soul—the two-dimensional, one-edged, and *one*-sided soul of a Möbius strip. This object is unusual because of its one-sidedness; most imaginable surfaces are all two-sided. Because of its having only one side, a Möbius strip could be entirely painted without the painter's ever having to cross an edge. It has other odd properties, too, including the one described in this limerick:

A mathematician confided
That a Möbius band is one-sided,
And you'll get quite a laugh
If you cut one in half,
For it stays in one piece when divided.

All together, topological results such as this portray, almost metaphysically, a complex but orderly geometric realm whose elements are as rationally interrelated as the living elements of the real world. As we've seen, topologists have learned that an object, like a living being, has an identity that is independent of its environment, an identity that is also indestructible against superficial changes in the object's appearance. Based as it is on this singular revelation, the entire body of topological knowledge implies that there is a scrutable order underlying the ostensibly chaotic variety of the geometric world.

This is what science has discovered about the physical world, but with one important difference: Whereas science is still at odds with most religions about the proper explanation for the observed orderliness of the physical universe, there can be less question that the orderliness of the geometric realm exists by design. In this case, the sole question is whether the design is of our own construction, or whether it is being revealed to us through our studies in topology.

In either case, our topological studies help us to comprehend not only the geometric realm but also the human realm. This is because in many ways topologists are farther along in their understanding of soul than others are. Topologists are able to tell us specifically what it means to be a strip of paper, or a basketball, but we are less well enlightened as to what it means to be human, judging from our differences of opinion.

This quandary isn't a strictly religious one, either; it should be of as much interest to an atheist as it is to a Jew. What is it about us individually that survives a lifetime of aging? And what is it about us collectively that might survive the millions of years of future evolutionary changes—if we are, in fact, evolving? These unresolved questions bear on the larger questions of our uniqueness and how we think of ourselves as fitting into the scheme of things; they also bear on the more modest question of how we can recognize old friends at reunions without resorting to name badges.

Mannerisms and personalities are properties and traits that, reminiscent of topological invariants, apparently remain little altered throughout a person's lifetime. But I wonder by what traits I might be able to distinguish humans from other animals if a reunion were held thousands of millions of years from now—a time period sufficiently long that, if Darwin was correct, we could expect some appearances to have changed dramatically.

Today, the criteria we rely on most to distinguish species of animals are based on a comparison of the animals' physical traits. Modern taxonomists have become quite good at classifying animals according to such things as the ratio of brain-to-body weight, skeletal construction, mode of locomotion, number of toes, and so on. We are less equipped to define species on the basis of more subtle criteria—a species's emotions, for instance. An animal's emotions are about as relevant to today's taxonomist as an object's number of edges is to the metric geometer.

Yet it is my conjecture that at that hypothetical reunion millions of years from now, it would be solely on the basis of such subtle criteria that I would stand any chance of recognizing my evolutionary kin. My topological hunch is that our quirks, and not our looks, are the relevant evolutionary invariants—the analogues of topological invariants. They are what it means now to be human and what it is likely to continue to mean throughout the course of evolutionary change.

Related Essays

Abstract Symmetry

Nothing Like Common Sense

Catastrophe Theory

THE FAMILIAR FACES OF CHANGE

Not Chaos-like, together crushed and bruised,
But, as the world harmoniously confused:
Where order in variety we see,
And where, though all things differ, all agree.

Alexander Pope

The only certainty in this world, as the saying goes, is change, and everything we learn from science bears this out. The contents of the universe and the universe itself are in an implacable state of flux. Things that appear constant—mountains, the atmosphere, the sun—are in fact constantly sustaining enormous changes of being; they are in an active state of balance. Even the cells in our own bodies are completely (and invisibly) replaced by new ones every seven years or so.

Our attitude toward change, especially abrupt change, has itself changed in the last several decades. In the past, things that happened suddenly, such as accidental deaths and natural disasters, were associated with some inscrutable agent such as a capricious god. The faces, the patterns, of abrupt change were mostly enigmatic to us and were therefore threatening. In the seventeenth century, during the Enlightenment, Isaac Newton did recognize that many instances of gradual change—population growth, for example—follow predictable patterns that can be represented by a few mathematical figures. But even then, and for decades thereafter, sudden changes appeared to be unsusceptible to such a mathematical categorization or to any satisfactory rational depiction at all.

About twenty years ago, the French mathematician René Thom did successfully categorize abrupt changes, or catastrophes, as he called them. He discovered that most catastrophes follow orderly patterns that are describable qualitatively in terms of seven mathematical figures. Though his discovery does not help us explain the origins of abrupt changes, it does indicate that these changes are not as irrational, as undisciplined, as they were previously believed to be.

It also indicates that the abrupt changes that affect us personally, such as nervous breakdowns and fits of spontaneous anger, are of the same qualitative variety as those sustained by the far-flung cosmic landscape. In this sense, Thom's theory enables us to recognize as never before the faces of catastrophic change that constantly transfigure the natural world.

In a dynamic world such as ours, it is inevitable that scientists require a mathematical means with which to study change, and it was mainly in response to this scientific need for a quantitative language of change that Newton invented the calculus. He designed it to describe changes that proceed in small, continuous steps—gradual changes, in other words—and so scientists use it for calculating such things as the motion of a planet around the sun, the growth of a population, or the constantly increasing speed of a falling object.

Just as significantly, however, they have learned in the process of using the calculus that gradually changing phenomena that appear to be unalike are actually related mathematically. For instance, the exponential-growth equation in calculus that describes the growth in value of a normal savings account is exactly the same one that describes the growth of bacteria in a petri dish, the growth in number of a chain letter, or the normal growth of an animal population.

Most notably, the calculus has enabled us to relate and to categorize all the possible motions of an object moving freely in a gravitational field. According to the calculus, such an object will follow one of only three archetypal trajectories, depending solely on the speed at which it is launched. Nothing about the projectile itself matters—not its shape, weight, density, or chemical composition. Irrespective of how different looking two projectiles may be, they will follow identical trajectories as long as they are launched at the same speed.

If a projectile is launched with a speed less than orbital speed (which for the earth's gravitational field is about 17,000 miles per hour), then its path will always resemble a parabola, the slowly curving arch usually followed by an arrow, a bullet, or a thrown rock. If, however, the object is launched with a speed between orbital speed and escape speed (which for the earth's gravitational field is about 25,000 miles per hour), then its path will always be an ellipse or a circle. Such is the case with all the planets orbiting the sun and with all the satellites circling the earth. And, finally, if an object is launched with a speed equal to or greater than escape speed, then its path will be a hyperbola, which looks like a parabola with its arch more severely bent. This is the path that was followed by American astronauts to the moon and by spacecraft traveling to other planets. The discovery of these categories was considered as revolutionary in its day as Thom's categorization of abrupt change is today.

For all its merits, however, the calculus still left scientists with no comparable mathematical means to describe and relate abrupt changes. For want of such means, biologists were hampered in their theoretical studies of cell division, one of the most basic phenomena in biology, since cells tend to divide suddenly rather than gradually. Apparently, Thom was not only aware of this problem in biology, he was in part motivated by it to develop his catastrophe theory; some of his earliest explications of the theory were published, in 1968, as part of a series of books entitled *Toward a Theoretical Biology*.

Scientifically and mathematically, catastrophe theory complements the calculus. Whereas the calculus is a quantitative theory of gradual change, catastrophe theory is a largely qualitative theory of abrupt change. Specifically, catastrophe theory is elaborated in the language of topology, the qualitative mathematical study of shapes. Topology itself is a branch of geometry that was founded in the eighteenth century by the Swiss mathematician Leonhard Euler.

One key notion of topology that is particularly evident in Thom's mathematical approach to abrupt change is the notion of topological equivalence. Two objects are said to be topologically equivalent if they share certain essential attributes, regardless of their other dissimilarities. For instance, a doughnut and a coffee cup are topologically equivalent because each has a hole in it. The hole is an essential feature in the sense that if we imagine transfiguring a doughnut into a cup or vice versa, almost everything else about the object will have changed in the process except for the hole; *it* persists. Topological equivalence is always judged on the basis of such immutable (qualitative) attributes only, not mutable (quantitative) details such as the size of the doughnut or the specific shape of the cup. For this reason, it is possible, and even common, for two very different looking objects to be topologically equivalent.

In developing catastrophe theory, Thom set out to find an analogous way of qualitatively relating catastrophes, even ones that might be as outwardly dissimilar as cups and doughnuts. For this reason, he sought an analogy to the hole, some essential attribute of catastrophes that might be used to mathematically describe and classify them.

Such an attribute, he discovered, is the number of factors

that control the dynamics of a catastrophe. By a control factor, Thom meant any element of a situation undergoing sudden change that can actually affect the progress and direction of that change. For instance, the essential factor controlling the inflation (and, ultimately, the catastrophic rupturing) of a balloon is air pressure. By reducing or increasing this one factor alone, the progress and direction of the balloon's changing status is affected.

This single discovery—that the number of control factors is as essential an attribute of a catastrophe as the hole is to a doughnut—is the basis of Thom's definition of catastrophic equivalence. According to the definition, two or more instances of abrupt change are catastrophically equivalent if each one's behavior is controlled by the same number of factors; the bursting balloon is thus catastrophically equivalent to any and every other abruptly changing phenomenon whose behavior is controlled by a single factor of some sort. Similarly, other catastrophes are grouped by Thom according to whether their behavior is controlled by two, three, or more factors.

This definition, Thom's analogy to topological equivalence, is the nub of catastrophe theory. Conceptually, it decrees that if two phenomena are controlled by the same number of factors, then they will follow the same qualitative pattern of change, even if they don't match quantitatively and in other details. Following this definition, therefore, Thom is able to recognize qualitative family resemblances among apparently unrelated catastrophes.

In a sense, Thom's discovery that the essential attribute of a catastrophe is the number of factors controlling its behavior is comparable to Newton's discovery that the essential attribute of a trajectory in a gravitational field is the projectile's launch speed. In both these domains, all other attributes of a changing situation except for the essential attribute are mathematically inconsequential. They are masks disguising faces of change that are, in some sense, fundamentally similar.

There are, however, differences between Thom's and Newton's findings. Whereas three categories alone cover the gravitational situation, no fewer than seven cover all of the more widely known instances of abrupt change in the natural world. Also, whereas the three representative trajectories exist in real space and are therefore easily visualized, the concrete representations of the seven archetypal catastrophes exist in an abstract, mathematical space and are not as easy to picture.

Perhaps the easiest way to visualize them is to imagine, as Thom did, that experiencing a catastrophic change is qualitatively like sustaining a precipitous fall while traveling on some hazardous geometrical surface—that is, in falling precipitously, as in changing abruptly, we suddenly end up in a place very different from that in which we were just before we fell. Technically, this analogy means that the mathematics of catastrophe theory is similar in some respects to the mathematics of geometrical surfaces, an established subject that is familiar to most mathematicians.

The factors that control the direction and progress of a catastrophe are imagined to control the direction and progress of our movements upon the corresponding abstract surface. This is why the surface is called the control surface, and the catastrophe is compared to falling off one part of the control surface and landing upon another. Also, it follows that there are as many dimensions to a catastrophe's control surface as there are control factors to the actual catastrophe: each factor controls movement along a single dimension.

If we do think of the faces of abrupt change as being like hazardous geometrical landscapes, then Thom's main discovery is that most of the landscapes we observe in nature and in ourselves involve only seven basic and distinct hazards. These are the seven familiar faces of change, and in an effort to help visualize them, Thom gave them names that are evocative of their shapes. In order of increasing complexity, the archetypal catastrophes are the fold, the cusp, the swallowtail, and the butterfly; there are three varieties of swallowtail and two of butterfly, thus bringing the total number to seven.

The fold catastrophe, which is controlled by a single factor, has the simplest face of all. Its control surface is technically not even a surface at all; it is a one-dimensional horizontal line that curves downward at one end to form the edge of a vertical cliff. According to catastrophe theory, this is the abstract mathematical visualization of an inflating balloon.

An increase in the balloon's air pressure corresponds graphically to movement along the line toward the cliff's edge, and a decrease corresponds to movement away from the cliff. Naturally, if we continue to inflate the balloon, a point is reached at which just one more molecule of air will be enough to blow the whole thing up. That point corresponds to being at the very edge of the cliff, and rupturing the balloon corresponds to falling off the edge. With a fold catastrophe, there is typically no bottom to the cliff, which means that once we've gone over the edge, there is no way back up. For our example, this signifies that once a balloon pops, it cannot be unpopped.

The normal aging process is also represented by a fold catastrophe, with the single control factor being time. Moving away from the cliff represents getting younger, and moving toward the cliff represents the normal situation, getting older. Our very last moment of being alive corresponds to standing on the cliff's brink; dying corresponds to falling off the cliff. According to catastrophe theory, therefore, a life follows the same mathematical pattern of change as an inflating balloon, and dying is qualitatively similar to a bursting balloon in the sense that once a life has been destroyed, it cannot be revivified.

This irreversibility is not true of a cusp catastrophe, which by definition is an instance of sudden change that is controlled by two factors. According to Thom's theory, a cusp catastrophe is distinguished by the way it can recover—either partially or fully—from the catastrophic change.

What he is referring to here can easily be seen from looking at the mathematical face of a cusp catastrophe. The central region of its two-dimensional control surface is dominated by an overhang whose shadow on the surface beneath it has a roughly triangular shape, like a cusp. Most notably, the overall layout is such that after one falls off the overhang, it is always possible to hike back up to the edge (all the while staying on the control surface).

The working of one of those little toy metal "clickers," or frogs, is a cusp catastrophe, because the behavior of a frog is controlled by two factors—in this case, finger pressure and the elasticity of its metal tongue. Graphically, the frog's tongue in the relaxed position corresponds to being in a region atop the cusp's overhang, far from the edge. Pressing on the tongue with increasing pressure corresponds to moving closer to the edge. At some point, the pressure is great enough that the metal tongue suddenly buckles, with a click; this corresponds to falling off the edge of the overhang, onto the control surface below. There it stays, in fact, as long as we hold the frog's tongue in its bent position. Letting go corresponds graphically to the frog's wending its way back up to the overhang.

Of the seven catastrophes, the cusp is the one whose face, whose mathematical pattern of change, we see most often in the world. Catastrophe theory was not designed to explain why this might be, but certainly any explanation would refer to the prevalence of opposites in reality and the episodes of reversible sudden change that relate them.

We recognize cusp catastrophes in some patterns of waking and sleeping, for instance. Normally, we fall asleep gradually, in stages, but sometimes the transition is more abrupt, such as after a hard workout. In such a case, it is always possible to recover—to wake up—from the catastrophic change, and this is the hallmark of a cusp catastrophe.

The cusp catastrophe is also a familiar face in the moody

behavior of a manic-depressive, in the spasmodic episodes of war and peace between irreconcilably hostile nations, and in the erratic highs and lows of the stock market. Like the metal frog, each of these examples of a cusp catastrophe involves a sudden swing from one extreme to another, followed by the possibility of returning to the original extreme. For this reason, unrelated though they may appear superficially, these phenomena can all be described with the same mathematical formulas and pictured in terms of the same catastrophic landscape. With catastrophe theory, therefore, we are able to see fewer differences in the world than we would without it.

The five remaining catastrophes are not as widely applicable as the cusp catastrophe; neither are their control surfaces as easily visualized. The problem with visualizing these surfaces is that they are all more than two-dimensional, in keeping with the greater number of control factors for these catastrophes, which makes them unlike what we would normally picture to be a surface. All we can do is imagine what their two-dimensional shadows must look like; in fact, the swallowtail and butterfly surfaces got their names from the imagined appearance of their 2-D silhouettes. (To show that the pictures are largely a matter of imagination, Thom was wont to point out that the swallowtail catastrophe was named by a blind colleague of his, the French mathematician Bernard Morin.)

Of the butterfly and swallowtail catastrophes, we see the face of the former more often in nature. Mathematically, the control surface of a butterfly catastrophe (or, more precisely, its 2-D silhouette) is distinguished by not one but two centrally located overhangs. Their relative arrangement is hierarchical, so that it is possible to fall off the topmost one and land either on the intermediate one or on the surface beneath them both. In either case, as with the cusp landscape, it is always possible to hike back up to the topmost overhang.

In most examples of butterfly catastrophes, the intermediate overhang represents a compromise between whatever is represented by the top and bottom levels of the control surface. For instance, two hostile nations that behave qualitatively like a butterfly catastrophe always have the option of negotiating rather than going directly from peace to war.

One of the few examples of butterfly catastrophes in which the intermediate overhang represents something more unusual than a compromise involves the affliction of mostly young women known as anorexia nervosa. Untreated anorexics behave like cusp catastrophes, flipping between episodes of fasting and gorging. Recently, however, the British mathematician E. C. Zeeman reasoned that if catastrophe theory is correct, the phenomenon of anorexia nervosa could be transformed from a cusp to a butterfly catastrophe by bringing additional control factors to bear.

Zeeman collaborated with a British psychotherapist, who devised a therapy for anorexics based on this hypothesis. The therapy involves putting the patient in a trance. Being put in a trance is a sudden change that corresponds to jumping from either the lower level (fasting) or the upper one (gorging) to the intermediate level of the butterfly catastrophe. It is at this median level that anorexics appear to be most receptive to psychological counseling. "When the patient is fasting," Zeeman ventures to explain, "she views the outer world with anxiety, and when she is gorging, she is overwhelmed by that world." However, he says, "during the trance [the patient] is isolated, her mind free both of food and of scheming to avoid food."

Although examples like this illustrate the scientific potential of catastrophe theory, there is presently a debate among scientists and mathematicians about how scientifically useful it will be in the long run. The pessimists generally argue that the scientific utility of the theory is limited by the qualitative nature of its founding notion, catastrophic equivalence. As we have seen, catastrophes controlled by the same number of factors are assumed in this theory to be equivalent, regardless of their detailed dissimilarities. Those mathematicians who are skeptical about the theory's scientific value point out that the very details that are ignored by the theory are those that science is liable to be interested in.

In the case of anorexia nervosa, for example, scientists undoubtedly wish to probe beyond the qualitative observations that the disorder follows an archetypal pattern of behavior that can be altered to follow another, different archetype; at some point scientists will also presumably wish to identify whatever detailed variations there might be to the archetypal pattern, in the hope of discovering which specific traits make a person more or less susceptible to becoming an anorexic.

However true such an argument might be, the qualitative character of Thom's theory is a welcome addition to mathematics and also to our culture's and age's accustomed way of looking at things. For one thing, the theory enriches the language of mathematics so that we can now express things about abrupt change that we were heretofore unable to express. In addition, Thom's theory and its qualitative foundation are also refreshing counterpoints to the reductionist's largely unchallenged reign of detailed, quantitative analysis. As the German physicist Bernhard Bavink once said, it allows us a rare opportunity to use mathematics "to place the concept of measurable and countable quantity in second place, and the basic biological concept of form, or gestalt, in first place."

To the extent that it *is* scientifically useful, Thom's theory bears out the vitalist's twin notions that the whole is greater than the sum of its parts and that looking at the whole can often be more enlightening than scrutinizing its parts. In defense of these notions, Thom lectures biologists and physicists at some length in his book *Structural Stability and Morphogenesis* on the dangers of looking at something too closely.

He challenges the common reductionist belief "that the interaction of a small number of elementary particles embraces all macroscopic phenomena when, in fact, the finer the investigation is, the more complicated the events are, leading eventually to a new world to be explained in which one cannot discern among the relevant factors for macroscopic order."

At the very least, catastrophe theory does enable us to recognize that, in cataloging change, there *is* some order in the multifarious changes that are constantly transfiguring the universe and its contents. Three hundred years ago, Newton found that a round, gold object in flight behaves basically no differently from, say, a square, silver one; the only factor that determines the trajectory of each through a gravitational field is the speed at which it is launched. Similarly, Thom has found that the sudden change experienced by a mother awakened by her child's cry follows the same mathematical pattern among all mothers, no matter what nationality or race they may be part of.

In Newton's case, the presence of a gravitational field accounted for the fact that all kinds of gradual motion belong to one of only three categories. Thom discusses the possible existence of a *champ vital* (life field) to account for the fact that sudden changes, too, appear to boil down to a few categories. He speculates, in fact, that the *champ vital* might be just like a "gravitational or electromagnetic field" in which all "living beings would then be particles . . . of this field."

Speculation of a more modest kind can lead us to see in Thom's theory an eloquent mathematical elaboration of a basis for our empathy with the natural world. By merely cataloging disparate phenomena, they become less mysterious and more familiar to us. I wonder whether the evidence of the presence here on earth of Thom's seven catastrophes extends outward from our solar system to other worlds, perhaps inhabited by other beings. If mathematics is the universal language many of us believe it to be, then sudden change anywhere else should have faces just as familiar as those on earth, though the possibility does exist for some queer juxtapositions. . . . Imagine a world where balloons click and frogs pop.

Related Essay

Locating the Vanishing Point

Combinatorics

OF WAR AND PEACE

> The problems of the world cannot possibly be solved by skeptics or cynics whose horizons are limited by the obvious realities. We need men who can dream of things that never were.
>
> John F. Kennedy

Comedian George Carlin once proposed a simple way of promoting worldwide peace: having everyone shake hands with and introduce themselves to everyone else. It's an engaging thought but, alas, pure fantasy—if each of us followed Carlin's advice, shaking a different hand every second, it would take about a hundred years to finish the task.

Carlin's joking proposal evidences a sober truth about human existence, namely, that many of our most precious fantasies are compromised by the reality of our limitations, and specifically our inability to manage large quantities. For example, in our world of four billion people, where a war can break out between any two or more persons, there is the potential for well over one hundred, billion, billion, billion different conflicts. With so many possible combinations for breaching the peace, we could not expect freedom from hostility even if every one of us spent some of our time being peacemakers.

In the world of mathematics, too, there are problems whose large combinations of elements overwhelm our means to resolve them. Representative of these is the age-old traveling salesman problem that turns on the query, In which order should a salesman arrange the cities he is supposed to visit so that he ends up traveling the least possible total distance? The answer is easy enough to fathom if only a few cities are involved, but for a large number of cities, say, fifty, there are more than ten thousand million, million different possible itineraries from which to choose.

Problems in mathematics that entail an analysis of combinations of things comprise the subject called combinatorics. The analysis usually consists of figuring out all possible combinations for the purpose of recognizing the one perfect combination that solves the problem.

Most such problems involve a manageable number of total possible combinations, so that a solution is readily picked out. This is the case with problems that require sorting through all the possible ways of pairing a dozen socks (66 possible solutions), forming fractions from the first ten numbers (45), or grouping a dozen guests at three separate tables (220).

In some problems, however, the number of combinations is so astronomical—more than the total number of stars in the entire universe—that not even the world's fastest computer can figure them all out in any reasonable amount of time. In particular, this is the case with the fifty-city traveling salesman problem. So daunting is the number of combinations in problems such as this that mathematicians' original fantasy of being able to solve them has been compromised to the more modest one of being able to work out *approximate* solutions that are optimal in only a less-than-perfect sense.

In combinatorics, it is customary to categorize a problem's degree of largeness according to how the time a computer needs to work out all possible combinations increases as the number of elements in the problem increases. (The number of elements is called the size of the problem, and in the traveling salesman problem it refers to the number of cities on the itinerary.) There are three degrees of largeness; in ascending order, these are called arithmetic, polynomial, and nonpolynomial.

With an arithmetic-class problem, the computer time needed to compute all possible combinations grows in simple proportion to the problem's size. This is the case with the computer-dating problem of searching through a pool of candidates to find someone who is most compatible with a particular client. If we double or triple the pool of candidates—that is, the size of the problem—we double or triple the necessary amount of computer time.

With polynomial-class problems, the required computer time grows more sharply, as some power (square, cube, and so on) of the problem's size. These include a problem such as figuring out how many different license plates can be made using three letters and three numbers, assuming that you have a pool of twenty-six letters and ten digits, zero through nine, to work from (Answer: 17,576,000). Doubling or tripling the pool of letters and digits increases the total possible combinations eightfold or twenty-sevenfold (that is, two cubed or three cubed), and the computer time needed to figure out the possibilities is increased accordingly.

For a nonpolynomial—or NP-class—problem, the necessary computer time grows exponentially with its size, following the same pattern of rapid growth as an increasing population of reproducing individuals. A large share of this class is comprised of scheduling problems in which the challenge is to identify the one schedule that produces the desired effect; these include the traveling salesman problem and the related problem of routing faced by the airline industry.

Airlines seek to schedule their stops so as to minimize expenses while providing as much service as necessary to as many cities as possible. The problem can easily get out of hand. In the case of one American airline, People's Express, the change from servicing a mere half dozen cities to some two dozen cities has meant that the number of possible routing schedules has grown from about thirty-two to 1,048,576. That's a 32,768-fold increase in the number of possible combinations corresponding to just a fourfold increase in the size of the problem. With such complexity, it is little wonder that at a typical airline it takes a staff of thirty people and two computers working for two years to prepare a single overall schedule.

Of the three classes of combinatorial problems, the arithmetic and polynomial ones grow slowly enough with the size of the problem that mathematicians can usually handle them with the aid of computers. Some of the smaller-sized NP-class problems, too, are manageable. But even with today's superfast computers, it would take about 30,000 years to work out all the possible combinations in a single NP-class problem the size of a fifty-city traveling salesman problem, and no computer in the foreseeable future will be able to solve large-sized NP-class problems either.

This last realization, which gradually sank in over the past couple of decades, is in stark contrast to the optimism mathematicians felt when the first digital computers appeared, back in the 1940s. With the UNIVACs came the belief that the fantasy of solving large-sized NP-class problems would become a reality as soon as a powerful enough computer could be built. As the years passed, computers did indeed become increasingly powerful, but as stated above, it became clear that they were not yet powerful enough to handle many large-sized NP-class problems and were not soon likely to be.

Rather than totally abandoning their fantasy of solving large NP-class problems, mathematicians chose to embrace a lesser goal. For roughly the last twenty years they have directed their attention to finding approximate solutions, using techniques that do not require finding all the possible combinations in a large-sized NP-class problem.

These optimizing techniques, as they are called, hinge on the mathematician's first being able to determine which criteria the actual solution is liable to satisfy. He then formulates a working solution that conforms to those criteria, much as a teacher formulates a curriculum that reflects some preconceived pedagogical criteria. Such a solution is said to be optimal because it meets the stated criteria, but unlike the actual solution, it is usually neither perfect nor unique. The simple reason for this is that the criteria from which the optimal solution is derived are neither of those things. Typically, in fact, they are little more than educated guesses.

One optimal solution for the large-sized traveling salesman problem is based on the rule of thumb that the salesman should always proceed from where he is to the nearest city. This is called the closest-unvisited-city criterion and it is obviously based on common sense.

Although this sensible criterion will not lead us to the actual solution, it usually comes within twenty percent of being perfect—that is, of producing the shortest possible itinerary. Mathematicians have arrived at this evaluation by comparing the optimal solution with the actual one for traveling salesman problems that are small enough to be solvable. The assumption, which is reasonable, is that this comparison is representative of the error that is incurred by having to settle for the optimal solution in the larger-sized problems.

The nonuniqueness of the optimizing criteria is particularly evident in the airlines' attempts to grapple with their NP-class scheduling problems. Currently, they use various combinations of four basic optimizing scenarios, called skip-stop scheduling, local scheduling, nonstop scheduling and cross-connection scheduling. All of them, to some degree, are derived simply from common sense.

In skip-stop scheduling, airlines fly directly to every other city along a route. This provides efficient service at intermediate distances. In local scheduling, airlines fly to every city along segments of routes, providing excellent short-range service. In nonstop scheduling, airlines fly directly to faraway cities, providing speedy long-distance service. And, finally, in cross-connection scheduling, airlines provide passengers with more destination options by terminating flights at a common check point, such as Chicago. The reasoning here is that no airline can possibly travel nonstop between every two cities. Therefore, a passenger flying from Providence, Rhode Island, to Santa Barbara, California, will be able to change planes at a heavily trafficked airport somewhere along the way.

Although the reality of the inability to compute large numbers of combinations has impelled most mathematicians to concentrate their efforts on formulating optimal solutions, a recent discovery in combinatorics has reawakened in some of them a bit of the old fantasy. In 1971, the American mathematician Richard Karp discovered that there is a subcategory of NP-class problems that are in some sense archetypes of the entire class. Karp called them NP-complete problems, because, as he proved, if mathematicians are ever able to solve just one NP-complete problem, they will then have the technical means to completely solve *all* NP-class problems. Included among these prototype problems is the traveling salesman problem.

Karp's revelation both has and has not brought mathematicians closer to solving large NP-class problems. That is, it has not ameliorated our inability to figure out large quantities of combinations, but it *has* focused the challenge for mathematicians considerably and in so doing has brought them a bit closer to a possible successful response.

At the same time, the direction in combinatorics today continues to be influenced as much by human nature as by anything else. By pursuing optimal solutions where actual solutions cannot be had, mathematicians are behaving like humans who will seek an approximate peace wherever a true peace is denied them, even as they persist in fantasizing about an end to conflict. And the comparison goes one important step further. Mathematicians have about as much chance of solving the fifty-city traveling salesman problem, it appears, as we do of securing worldwide peace. And yet, human nature is such that Karp's result in combinatorics and George Carlin's proposal for the real world can evoke in us the very same sense of hope, however remote that hope is. The hope is that if we can solve but a single key problem of the teeming masses, be it the traveling salesman problem or the morass of human conflict, all problems related to it will thereupon be resolved.

CHRONOLOGY

Here is a list of the mathematicians referred to in the essays, along with the years in which they lived and brief descriptions of their contributions to the history of mathematics.

B.C.

624–546 THALES OF MILETUS is generally credited with having introduced the concept of logical proof into geometry.

569–500 PYTHAGORAS OF SAMOS made many contributions to the early history of geometry, including the Pythagorean theorem. He and his school of followers also developed the basis of modern numerology and discovered irrational numbers.

429–348 PLATO, a pupil of Socrates, formulated the philosophy of ideal forms, which influences today's Platonist philosophy of mathematics.

384–322 ARISTOTLE is the author of the *Organon,* the seminal work on traditional deductive logic.

330–275 EUCLID, a mathematics tutor in ancient Alexandria, compiled and reformulated the results of Greek geometry into a single, cohesive, logical subject. The result was a voluminous treatise entitled the *Elements,* which is the world's second all-time best-seller, behind the Bible.

287–212 ARCHIMEDES developed a numerical system that enabled us, for the first time, to express the size of the universe. He also made important contributions to the mathematics that antedated the differential calculus.

A.D.

100–168 CLAUDIUS PTOLEMY is especially remembered for his brilliant use of mathematics in astronomy.

1596–1650 RENÉ DESCARTES (French) re-expressed the results of geometry in a novel and productive way, thereby creating the subject we now call analytic geometry.

1642–1727 SIR ISAAC NEWTON (English) developed, independently of Leibnitz, the differential calculus.

1646–1716 BARON GOTTFRIED WILHELM VON LEIBNITZ (German) developed the differential calculus, independently of Newton.

1707–1783 LEONHARD EULER (Swiss) is credited with the founding of topology.

1777–1855 CARL FRIEDRICH GAUSS (German) was an all-around genius, contributing to nearly every area of modern mathematics. In particular, he founded the subject of differential geometry.

1790–1868 AUGUST FERDINAND MÖBIUS (German) contributed to the early development of topology.

1793–1856 NICHOLAS IVANOVICH LOBACHEVSKI (Russian) developed, independently of Bolyai and Gauss, a non-Euclidean geometry.

1802–1860 JÁNOS BOLYAI (Hungarian) developed, independently of Lobachevski and Gauss, a non-Euclidean geometry.

1811–1832 ÉVARISTE GALOIS (French) made founding contributions to the subject called group theory.

1814–1897 JAMES J. SYLVESTER (English) contributed, with Cayley, to the development of abstract algebra through the invention of matrices.

1821–1895 ARTHUR CAYLEY (English), with Sylvester, invented matrices.

1826–1866 GEORGE FRIEDRICH BERNHARD RIEMANN (German) contributed to the modern concept of dimension with his embellishments of differential geometry, a subject created by Gauss.

1831–1916 JULIUS WILHELM RICHARD DEDEKIND (German) clarified the definition of irrational numbers.

1845–1918 GEORG CANTOR (German), creator of classical set theory, contributed to our modern understanding of infinity.

1848–1925 GOTTLOB FREGE (German) formulated arithmetic in terms of logic and set theory, the basis of today's "new math."

1849–1925 FELIX KLEIN (German) organized our understanding of geometry by classifying its various forms.

1854–1912 JULES HENRI POINCARÉ (French) contributed to all branches of modern mathematics.

1862–1943 DAVID HILBERT (German) made many contributions to modern mathematics and to its philosophy.

1872–1970 BERTRAND RUSSELL (English) was a major contributor to the fields of logic and the philosophy of mathematics.

1906–1978 KURT GÖDEL (German) is known for his proof that there will always be mathematical statements that are true but that cannot be proved true using logic.

1922–1974 IMRE LAKATOS (Hungarian) incorporated Gödel's result into a philosophy of mathematics called the philosophy of dubitability.

GLOSSARY

These are some of the more important mathematical terms that are used in the essays, along with brief definitions.

Algebra (*n.*) A generalization of arithmetic involving the addition, subtraction, multiplication, and division of many more kinds of numbers than just positive numbers.

Arithmetic (*n.*) The study of how positive numbers are added, subtracted, multiplied, and divided.

Asymptote (*n.*) In geometry, the boundary line toward which another line may approach with ever-increasing proximity, yet never actually meet.

Average (*n.*) In statistics, some trait that typifies the members of a group.

Axiom (*n.*) See *Postulate*.

Catastrophe Theory (*n.*) The mathematical study and description of abrupt change.

Chaos (*n.*) In statistics, chaotic behavior is behavior that is completely unpredictable.

Combinatorics (*n.*) The study of problems that involve combinations of elements.

Commutative (*adj.*) In algebra, refers to the fact that it doesn't matter in which order you add or multiply two numbers; you get the same answer either way.

Complex Plane (*n.*) In algebra, a mathematical map on which the coordinates of each point correspond to one real number and one imaginary number.

Conjecture (*n.*) An unproved, educated guess at some mathematical truism (for example, Goldbach's conjecture).

Continuum Hypothesis (*n.*) The conjecture, proposed by Georg Cantor in 1878, that there are an $\aleph_1$ number of irrational numbers (see *Transfinite Numbers*).

Correlation (*n.*) In statistics, a measured interdependence between two or more traits (for example, cigarette smoking and lung cancer).

Differential Calculus (*n.*) The mathematical study and description of change that is smooth and continuous.

Empty Set (*n.*) See *Null set.*

Exponential Growth (*n.*) In reference to a population, describes growth that increases at a rate that is always proportional to the size of the existing population. In other words, the *larger* the population becomes, the *faster* it grows.

Fractal (*n.*) An object whose number of dimensions is a fraction.

Fraction (*n.*) One whole number divided by another whole number (for example, 3/2, 5/3, and so forth).

Game Theory (*n.*) The study of conflicts of interest between two or more persons.

Geometry (*n.*) The study of the shapes and sizes of things, carried out in any one of several different ways. For instance, analytic geometry partakes of algebraic techniques to carry out a study of shape and size, whereas differential geometry uses the techniques of differential calculus.

Goldbach's Conjecture (*n.*) The conjecture that every even number can be expressed as the sum of two prime numbers (for example, 4 equals 3 + 1 and 8 equals 3 + 5).

Group Theory (*n.*) A branch of mathematics that enables us to describe symmetries that are imperceptible to the senses.

Homeomorphism (*n.*) See *Topological Transformation.*

Hypothesis (*n.*) See *Conjecture.*

Imaginary Number (*n.*) In algebra, the name given to a number whose square is a negative number. (Compare *Real number*).

Invariant (*n.*) In topology, any property of an object that survives a topological transformation.

Irrational Number (*n.*) A number that cannot be expressed as a whole number or a fraction, but only as an infinitely long decimal (for example, the number pi, which is 3.14159 . . .).

Manifold (*n.*) In topology, a conceptual universe having any number of dimensions.

Möbius Strip (*n.*) In topology, the one-sided band that is made by giving one end of a strip of paper a half twist before gluing it to the other end.

Negative Number (*n.*) In algebra, a number that is less than zero.

Null Set (*n.*) The set with no members.

Number Line (*n.*) A line along which the whole numbers, fractions and irrational numbers are arranged in increasing order.

Odds (*n.*) See *Probability*.

Optimal Strategy (*n.*) In two-person game theory, the strategy that guarantees a player either a win or a tie.

Paradox (*n.*) An argument that usually either leads to a dilemma or contradicts common sense.

Parallel (*adj.*) In geometry, it describes two or more lines that never cross, no matter how long they are extended.

Philosophy (*n.*) In mathematics, the study concerned with subjects such as the nature of mathematical existence and mathematical proof.

Pi (*n.*) The sixteenth letter of the Greek alphabet and the name given to the ratio of a circle's circumference to its diameter—roughly equal to 3.14159.

Point (*n.*) One of the basic elements of geometry; an entity that is imagined to have position but not dimension.

Polygon (*n.*) In geometry, a 2-D, many-sided figure.

Positive Number (*n.*) A number that is greater than zero.

Postulate (*n.*) A statement that is assumed to be true without proof

Prime Number (*n.*) A number that can be cleanly divided by only 1 and itself (for example, 1, 3, 5, 7, 11, 13, and so forth).

Probability (*n.*) A number (usually a percentage) expressing the likelihood of some one thing happening out of many possibilities.

Proof (*n.*) A logical argument that establishes the truth of a hypothesis.

Pseudosphere (*n.*) In geometry, an object whose shape resembles that of two heralder's trumpets with their bells butted together.

Pythagorean Theorem (*n.*) In geometry, a formula that enables one to calculate the length of the third side of a square-cornered triangle if the lengths of the other two sides are known.

Rational Number (*n.*) A number that can be expressed as the ratio of two whole numbers.

Real Number (*n.*) A number whose square is a positive number.

Sample (*n.*) In statistics, a group that is representative of the population from which its members are taken.

Set (*n.*) In set theory, any collection of real or imagined objects.

Singularity (*n.*) A mathematical point at which some quantity is infinite.

Solid (*n.*) In geometry, any object having three dimensions.

Square (*n.*) In arithmetic, equal to a number multiplied by itself (for example, 9 is equal to the square of 3).

Square Root (*n.*) In arithmetic, it is equal to that number that when multiplied by itself will produce a given number (for example, 3 is the square root of 9).

Syllogism (*n.*) In logic, a statement that usually consists of two postulates and a conclusion.

Theorem (*n.*) A mathematical conclusion that has been proved according to the rules of some system of logic.

Topology (*n.*) A branch of geometry that studies those properties of an object that remain unchanged throughout a topological transformation.

Topological Transformation (*n.*) The subjecting of an object to bending, twisting, and stretching.

Transfinite Numbers (*n.*) The hierarchy of infinities defined by Georg Cantor in the nineteenth century and designated by the succession of symbols $\aleph_0$, $\aleph_1$, $\aleph_2$, $\aleph_3$, et cetera (where $\aleph$ is aleph, the first letter in the Hebrew alphabet). Each aleph is infinitely larger than the one immediately preceding it, and $\aleph_0$ corresponds to the infinity we ordinarily speak of when we say "infinity."

Whole Number (*n.*) Any one of the counting numbers: 0, 1, 2, 3, 4, and so forth.

SUGGESTED READING

I recommend the following publications for their clear and engaging presentations of material that is referred to in these essays, and then some. In most cases, each publication has an extensive bibliography that will guide you even further along in your readings. I have selected these particular publications, too, because they are liable to be in most community libraries.

Historical

Bell, E.T. *The Development of Mathematics* (McGraw-Hill, New York, 1949).

Sanford, Vera. *A Short History of Mathematics* (Houghton Mifflin, Boston, 1930).

Struik, Dirk J. *A Concise History of Mathematics* (Dover, New York, 1967).

Biographical

Bell, E.T. *Men of Mathematics* (Simon and Schuster, New York, 1961).

Turnbull, H.W. *The Great Mathematicians* (NYU Press, New York, 1961).

SUGGESTED READING

General

Bergamini, David. *Mathematics* (Time-Life Science Library, Time, Inc., New York, 1963).

Fadiman, Clifton, ed. *Fantasia Mathematica* (Simon and Schuster, New York, 1958).

Fadiman, Clifton, ed. *Mathematical Magpie* (Simon and Schuster, New York, 1964).

Kasner, Edward and Newman, James. *Mathematics and the Imagination* (Simon and Schuster, New York, 1962).

Kline, Morris, ed. *Mathematics in the Modern World* (Freeman, San Francisco, 1968).

Kline, Morris, ed. *Mathematics: An Introduction to its Spirit and Use* (Freeman, San Francisco, 1978).

Pedoe, Dan. *The Gentle Art of Mathematics* (Dover, New York, 1972).

Smith, David E. *The Poetry of Mathematics and Other Essays* (Scripta Mathematica, New York, 1947).

Topical

Adler, Alfred. "Mathematics and Creativity" (*The New Yorker,* February 19, 1972).

Dieudonné, J. "Should We Teach Modern Mathematics?" (*American Scientist,* January/February 1973).

Kolata, Gina Bari. "Mathematical Proof: The Genesis of Reasonable Doubt" (*Science,* June 4, 1976).

Steen, Lynn Arthur. "Order from Chaos" (*Science News,* May 3, 1975).

INDEX